SpringerBriefs in Environment, Security, Development and Peace

Mediterranean Studies

Volume 4

Series Editor

Hans Günter Brauch

For further volumes:
http://www.springer.com/series/11792
http://www.afes-press-books.de/html/SpringerBriefs-ESDP-MeS.htm

Gamal M. Selim

Global and Regional Approaches to Arms Control in the Middle East

A Critical Assessment from the Arab World

Springer

Gamal M. Selim
Department of Political Science
Port Said University
Port Said
Egypt

ISSN 2193-3162 ISSN 2193-3170 (electronic)
ISBN 978-3-642-29313-9 ISBN 978-3-642-29314-6 (eBook)
DOI 10.1007/978-3-642-29314-6
Springer Heidelberg New York Dordrecht London

Library of Congress Control Number: 2013938167

Title Page Illustration: The title page photo is based on a combination of two sources that are available on these Websites: http://www.un.org/Depts/Cartographic/map/profile/mideastr.pdf and of http://lewis.armscontrolwonk.com/archive/2000/buff-blogging-for-arms-control. There is no source for the photo of the military airplane provided. It was obviously taken from the website of the US Department of Defense which is in the public domain.

Printed on acid-free paper

Springer is part of Springer Science+Business Media (www.springer.com)

Acknowledgment

I would like to thank the Simon Foundation in Canada and the International Security Research and Outreach Programme [ISROP] of Foreign Affairs and International Trade Canada [DFAIT] for supporting an earlier draft of this study in 2009.

Contents

Abbreviations

ABM	Anti-Ballistic Missile
ACPSS	Al-Ahram Center for Political and Strategic Studies
ACRPS	Arab Center for Research and Policy Studies
ACRS	Arms Control and Regional Security
AIPAC	American-Israeli Public Affairs Committee
AU	African Union
BWC	Biological Weapons Convention
CBMs	Confidence-Building Measures
CFE	Treaty on Conventional Armed Forces in Europe
CIA	Central Intelligence Agency
COARM	Working Group on Conventional Arms Exports
CSBMs	Confidence and Security Building Measures
CSCE	Conference on Security and Cooperation in Europe
CSCM	Conference on Security and Co-operation in the Mediterranean
CTBT	Comprehensive Test Ban Treaty
CWC	Chemical Weapons Convention
DCI	Defense Counterproliferation Initiative
DOP	Declaration of Principles
EU	European Union
EMP	Euro-Mediterranean Partnership
GCC	Gulf Cooperation Council
IAEA	International Atomic Energy Agency
ICBM	Intercontinental Ballistic Missile
INF	Intermediate-Range Nuclear Forces Treaty
JMP	Joint Meeting Parties
MAD	Mutually Assured Destruction
MTCR	Missile Technology Control Regime
NAC	North Atlantic Council
NATO	North Atlantic Treaty Organization
NPR	Nuclear Posture Review
NPT	Nuclear Non-Proliferation Treaty
NSCWMD	National Strategy to Combat Weapons of Mass Destruction
NSS	National Security Strategy

OSCE	Organization for Security and Co-operation in Europe
PBMs	Partnership-Building Measures
PLO	Palestinian Liberation Organization
PNAC	Project for the New American Century
PSI	Proliferation Security Initiative
QIZ	Qualified Industrial Zone
SALT	Strategic Arms Limitation Talks (Treaty)
SCAF	Supreme Council of the Armed Forces
SLBM	Submarine-Launched Ballistic Missile
START	Strategic Arms Reduction Talks (Treaty)
TNC	Transitional National Council
UAE	United Arab Emirates
UN	United Nations
UNEF	United Nations Emergency Forces
WMD	Weapons of Mass Destruction
WMDFZ	Weapons of Mass Destruction Free Zone

Chapter 1
Introduction

The end of the Cold War saw a return of the concept of arms control to the forefront of the academic and policy-oriented discourse on security. At the academic level, the end of the Cold War brought new conceptualizations of arms control. While the traditional neo-realist approach viewed arms control as an instrument for managing the balance of power between states and adjusting their military capabilities, the neo-liberal institutionalist approach conceptualized arms control as an instrument that could help to shape political perceptions of states, remove their security dilemma, and contribute to conflict prevention and possibly conflict resolution. At the policy-oriented level, the concept of arms control also gained considerable importance as an important mechanism for conflict resolution and security. In this context, Western powers have frequently advocated arms control as a conflict resolution strategy in conflict-ridden regions, such as the Middle East, South Asia and the Korean peninsula.

In the Middle East, the question of arms control also moved to the forefront of the region's politico-security agenda in the post-Cold War era. The end of the Cold War, in combination with other profound developments at the global and regional levels, resulted in renewed Western interest in arms control after decades of modest efforts with meager results. In this context, Western powers gave arms control a prominent position in the Arab–Israeli peace process which began at the Madrid peace conference in October 1991, with arguments that arms control would be an important prelude to the settlement of the Arab–Israeli conflict. During this period, it was common for these powers to approach the question of arms control in the Middle East from an international regime perspective. This was reflected in the formulation of a number of global arms control proposals for the purpose of establishing an arms control regime in the Middle East, such as the US Arms Control Plan in the Middle East, the French Arms Control and Disarmament Plan, and the *Arms Control and Regional Security* (ACRS) working group. In return, these proposals, among others, had the impact of triggering an intensive arms control debate in the region as they received mixed reactions from regional actors.

G. M. Selim, *Global and Regional Approaches to Arms Control in the Middle East*,
SpringerBriefs in Environment, Security, Development and Peace 4,
DOI: 10.1007/978-3-642-29314-6_1, © The Author(s) 2013

With the beginning of the new millennium, the Western and particularly American approach to arms control witnessed a major transformation as a result of a number of developments; chiefly among them were the coming into power in Washington of a neo-conservative administration, followed by the 9/11 attacks on New York and Washington. The new American approach shifted away from the traditional emphasis on cooperative, multilateral arms control in the direction of selective unilateral disarmament. This new thinking manifested in the rise of an American foreign policy doctrine, commonly known as the Bush doctrine, which replaced the principles of mutual deterrence, international regimes and collective security with those of unilateralism and pre-emption as the cornerstones of American national security strategy after 9/11. In the field of arms control, the Bush doctrine supported measures of proactive counter-proliferation and possible 'regime change' against states suspected of developing *weapons of mass destruction* (WMD). In fact, the Middle East became the main testing ground for the Bush doctrine. This was clearly evident in the case of the American invasion of Iraq under the justification of destroying Iraqi WMD.

This book aims to analyze the dynamics of arms control in the Middle East in the post-Cold War era. In this context, the book targets three major tasks. First, it examines global and regional arms control projects in the Middle East following the end of the Cold War and their impact on arms control efforts in the region. This is approached with a view to exploring aspects of continuity and change in the global arms control approach in the Middle East over the past two decades. Second, the book assesses Arab perceptions of the motivations for and constraints on establishing arms control regimes in the region. The analysis of Arab perceptions is also conducted with a view to delineating patterns of continuity and change in Arab approaches to arms control. Finally, the book explores the prospects of regional arms control in the wake of the Arab revolutionary wave which began in Tunisia in December 2010 and extended later to Egypt and other Arab countries. The Arab revolutions, commonly known as the Arab Spring, are expected to have significant ramifications for Arab domestic and regional politics, including the prospects of arms control in the region.

The book adds new insights to the literature by offering an Arab perspective on the arms control debate in the Middle East. Indeed, if one reviews the arms control literature on the Middle East, one will find that the majority of the literature has been written by Western, and to a lesser extent Israeli, academic and policy-oriented experts in the field. Although a number of Arab academics addressed the subject in their studies, such as those conducted by Al-Ahram Center for Political and Strategic Studies (ACPSS) in Cairo, the majority of these studies were written in Arabic and were largely atheoretical, and did not appear therefore to have achieved significant recognition beyond the local community of Arab social scientists. In this context, the emphasis in this book on Arab perceptions enriches contemporary debate by providing a new perspective on the arms control discourse, which has up till now been dominated by Western perspectives.

The definition of the Middle East is problematic. The term entails an arbitrary definition of a region which includes countries with little commonality in their political, social and cultural backgrounds. It has been used to delineate the boundaries of a region extending beyond the Arab world to include, in addition to the Arab countries, other non-Arab countries such as Pakistan, Turkey, Afghanistan, Iran, and later Israel. Historically, the term 'Middle East' did not have any definite geographical boundary, and was viewed as an artificial abstraction that was invented by European imperialist powers, mainly Britain, to serve their interests against a rapidly growing wave of Arab nationalism. In facts, Arabs themselves never referred to their region as the Middle East until the colonial usage of the term became current and stuck. As explained by Ismael (2011, p. 1), the meaning of the Middle East "has been determined by political rather than geographical factors and therefore has changed in correspondence with the growth of Western interest and involvement in the area… [the term] was invented as a formulation of British security discourse…[and] gained wide circulation during the Second World War when the Middle East Supply Centre was established by the United States and Britain."

During the Cold War era, the United States, alongside Britain, got more closely involved in the region, and the term 'Middle East' further became integral to American political and security discourse. This manifested in the rise of a series of American-led projects that sought to restructure the political and security map of the Arab world in order to align it with Western strategic interests. Perhaps the most important was the Middle Eastern project of the early 1950s, which (i) viewed the Middle East as a geographical area, rich in oil resources, containing a mosaic of nations, and threatened by external hostile powers, mainly the Soviet Union; and (ii) sought to link the region to the West through a security regime and a set of economic arrangements that would stabilize the region and preserve Western interests.[1] In the post-Cold War era, several Western attempts were also made to revive the Middle Eastern project in the Arab world. These included (i) the Madrid Arab–Israeli peace conference, held in October 1991. The conference branched off into two tracks; a bilateral and a multilateral track. Under the multilateral track, five working groups were formed and many non-Middle Eastern countries took part in them. In fact, the working groups were chaired by big powers and their meetings were held in different places in and outside the Middle East in order to emphasize the "internationalization" of the multilateral track; and (ii) the *Middle East and North Africa* (MENA) series of conferences held to formulate a pan-Middle Eastern regime for economic cooperation and to integrate Israel into

[1] The United States first set out the project in 1950 when it suggested establishing a *Middle East Defense Organization* (MEDO), and President Eisenhower reintroduced the idea in 1953 when he proposed to establish an alliance between the northern tier countries of the Middle East in order to contain the Soviet Union. The countries meant were Turkey, Iran, Pakistan and Iraq. In response to this proposal, Iraq and Turkey signed a security pact that became known as the Baghdad Pact in February 1955, and Britain joined in April 1995; under this security pact the three countries pledged to defend each other in case of a foreign aggression. The Pact, however, was met with vehement opposition from Egypt and advocates of the Arab regional system, and this led to its collapse in 1958 following the outbreak of the Iraqi revolution.

the region. The dimensions of the new Middle Eastern project were revealed by Shimon Peres, the then foreign minister of Israel, with the publication of his famous book *The New Middle East* in 1993 in which he envisioned a new regional cooperation system.

Indeed, this led many Arab scholars to dismiss the concept of the 'Middle East' as a Western invention and an attempt to undermine the Arab identity of the region by extending it to include other non-Arab actors. Accordingly, the concept of an Arab regional system was substituted, which represents more truly the interactions and relationships in the area. According to this view, the region is characterized by the dominance of one Arab nation with common features, interests, and security concerns and priorities. The Arab peoples, it is argued, enjoy a unity of language and culture by which they are entitled to form their own security and economic arrangements. Unlike the Middle Eastern system envisioned by the USA, which was based on the assumption that the Soviet Union was the main source of threat to the region, the idea of the Arab regional system was based on the assumption that the main source of threat was Israel, and that the defence of the region should be placed within an Arab security framework. This framework was represented in the establishment of the League of Arab States in 1945 and the conclusion of the Arab Common Defence and Economic Cooperation Treaty in 1950 (Riad 1981; Matar and Dessouki 1983; Al-Mesiri 2006).

Nevertheless, as a modern conceptual and geographic unit, the delineation of the Middle East followed in this book includes the countries in the Northern Belt (Turkey and Iran); the Fertile Crescent (Iraq, Syria and Lebanon, Israel and Palestine); those west and east of the Red Sea (Egypt and the members of the Gulf Cooperation Council); and the Arab Maghrib states of North Africa.

The organization of this book reflects the effort to set the question of arms control in the Middle East in the context of patterns of continuity and change throughout the region. Chapter 2 presents the theoretical foundation of arms control and its impact on security. It delineates the boundaries of the term, examines contemporary research programmes, and evaluates their contribution for understanding the dynamics of arms control and its applications between states. Chapter 3 addresses the rise of the arms control debate in the Middle East in the 1990s. It outlines both global and regional arms control projects during this era. Chapter 4 examines the status of arms control in the Middle East in the post 9/11 era. Chapter 5 delineates Arab perceptions of arms control efforts from the early 1990s up until the outbreak of the Arab Spring uprisings. Chapter 6 examines the prospects of regional arms control, in addition to Arab perceptions, in the aftermath of the Arab Spring. These perceptions will be examined at the governmental and non-governmental levels, with a view to comparing them, identifying elements of change and continuity and assessing their implications for the prospects of arms control in the Middle East. This review is based on the assumption that perceptions influence the prospects of regional stability and conflict resolution. Chapter 7 summarizes the main findings and outlines policy recommendations as far as arms control efforts in the Middle East are concerned.

References

Al-Mesiri, Abdel-Wahab (2006). The New Middle East in the American-Zionist Vision. *Al-Jazeera*, 1 November; available at: http://www.aljazeera.net/opinions/pages/6deae7db-b26a-42d0-b95f-b8a758e1c2cd

Ismael, T. Y., & Ismael, J. S. (2011). *Government and politics of the contemporary Middle East: Continuity and change*. London: Routledge.

Matar, J., & Dessouki, A. E. H. (1983). *The Arab regional system*. Cairo: Center for Arab Unity Studies.

Riad, M. R. (1981). A view from Cairo. In M. Ayoob (Ed.), *The Middle East in world politics*. London: Croom Helm.

Chapter 2
Arms Control and Security Cooperation: Contending Approaches

Arms control is but one of a series of alternative approaches to achieving international security through military strategies. Although the basic idea of arms control has its roots in the nineteenth century, the rise of modern arms control as a theory and practice can be traced to the Cold War era as an outcome of the American-Soviet nuclear arms race. In fact, arms control started to assume considerable importance in the field of security studies toward the late 1960s when the two superpowers entered their *Strategic Arms Limitation Talks* (SALT) in Vienna and Helsinki in 1969 and concluded their first arms control agreement, SALT I, in 1972. Since then, the Americans, the Soviets and the Europeans have spent more than 30 years in discussing, negotiating, and signing different agreements on arms control in both the nuclear and the conventional fields.

It is important to distinguish arms control from disarmament. Although the two concepts might share some commonalities, one must treat them as two distinct terms with different assumptions and working mechanisms. In general terms, disarmament entails the elimination of certain classes of weapons from the arsenals of states. The *United Nations* (UN) General Assembly has defined the term as "the elimination of all WMD", coupled with the "balanced reduction of armed forces and conventional armaments, based on the principle of undiminished security of the parties with a view to promoting or enhancing stability at a lower military level".[1] In return, the purpose of arms control is mainly regulatory, as it tends to put certain limitations on the acquisition, production, deployment and use of weapons. More specifically, arms control tends to "ban certain classes of weapons and weapons systems, place upper limits on the number of weapons that states may posses, limit the size and destructive power of weapons, ban the production of weapons that will increase the likelihood of war, and stop or at least slow the development of new technologies" (Griffiths and O'Callaghan 2002: 6). In *The*

[1] "Strengthening International Regimes for Arms Control and Disarmament". Background Note prepared by the United Nations Department for Disarmament Affairs, New York, 19–20 October 2004.

G. M. Selim, *Global and Regional Approaches to Arms Control in the Middle East*, SpringerBriefs in Environment, Security, Development and Peace 4,
DOI: 10.1007/978-3-642-29314-6_2, © The Author(s) 2013

Control of the Arms Race, Bull (1965: 7), a classic arms control scholar, defined arms control as restraints internationally exercised on the development, use and employment of weapons. Booth (1987: 140–142) subscribed to the same view. He argued that whereas disarmament is revolutionary in focus as it is based on upturning the traditional processes of military security, arms control is more conservative in focus as it seeks to regulate such processes. Accordingly, the main objective of arms control is regulation rather than elimination of weapons systems. In fact, arms control can lead states to agree to increases in certain categories of armaments if such increases will contribute to crisis stability, and thereby reduce the chances of war.

In practice, arms control was devised during the Cold War period as an alternative to disarmament, which for many had fallen into discredit as a means of reducing the likelihood of war. The German disarmament experience is a case in point. Although Germany had been forced to disarm following World War I, this did not prevent it from becoming belligerent again, nor did it restrict its ability to go to war during the 1930s. In the early post-war period, the United States and the Soviet Union held a series of disarmament negotiations under the auspices of the UN. However, such negotiations did not go beyond formal propaganda, as the main focus of the superpowers became the arms race rather than disarmament. The turning point came in the mid-1950s when after years of discussion on the reduction of armaments and the elimination of nuclear weapons, the United States backtracked on its previous commitments to seek disarmament arrangements, arguing that "the advances in modern armaments, including nuclear weapons, have been so significant that much of the earlier discussions of the inspection and control problems may well be outmoded".[2] The failure of disarmament negotiations to produce tangible results led eventually to the rise of a new thinking in the academic and policy circles concerned with the implication of the American–Soviet nuclear arms race. The new thinking replaced disarmament as an immediate goal with limited partial measures that would control the arms race and military power rather than eliminate them, since it held that elimination had proved to be unrealistic and even dangerous, and would not necessarily reduce the likelihood of war. In this context, whereas advocates of disarmament had formerly seen it as an alternative to military strength, arms control was now viewed as an integral part of military power, as its advocates sought to create a stable balance of power in which the forces that cause states to go to war could be controlled and regulated (Larsen 2002: 5–6).

Some analysts have also distinguished arms control from *confidence-building measures* (CBMs). The concept of CBMs refers to "collective arrangements about the function and use of military power in peacetime…designed to confirm non-aggressive intentions of all states and therefore build stable expectations

[2] Statement by Harold E. Stassen to the UN Disarmament Subcommittee, 6 September 1955. *Documents on Disarmament 1945–1959*, Vol. I. (Washington, DC: ACDA, 1960), p. 512.

concerning their military activities as instrument that rather deal with issues of military intensions and (mis)perceptions" (Rittberger et al. 1990: 70). This involves "the communication of credible evidence of the absence of feared threats by reducing uncertainties and by constraining opportunities for exerting pressure through military activities" (Holst and Melander 1977: 147). From an operational perspective, CBMs include measures such as (i) communication measures, i.e. the establishment of hotlines between the political and military leaderships; (ii) information measures, including the annual exchange of information on military forces, major weapon and equipment systems, and military budgets; (iii) notification measures, including the notification in advance of military manoeuvres and troop movements; (iv) observation measures, including the invitation of observers to major military manoeuvres; (v) compliance and verification measures, such as on-site inspections; and (vi) constraint measures, such as the establishment of demilitarized zones (Reich 1994: 240; Gunduz 1994: 188).

Although the very basic idea of confidence-building has its roots in the seventeenth-century peace of Westphalia and in many other subsequent agreements over the eighteenth and nineteenth centuries, the rise of CBMs as a modern concept was officially endorsed in 1975 with the convening of the first meeting of the *Conference on Security and Cooperation in Europe* (CSCE). The CSCE meeting adopted the Helsinki Final Act, which produced *The Document on Confidence-Building Measures and Certain Aspects of Security and Disarmament*. This was followed by the development of several generations of CBMs within the framework of the CSCE process.

Some scholars view CBMs as different from arms control. According to Holst (1987: 31), a classic CBMs analyst, arms control and CBMs are two separate concepts that adhere to different assumptions and mechanisms. He explained that CBMs "do not constitute substitutes for arms control, but they can pave the way for arms control and broaden and reinforce recognition of shared interests in the avoidance of war". This distinction is valid when the concept of arms control is narrowly defined as an instrument that deals only with the actual reduction of armament. However, the line between arms control and CBMs becomes difficult to draw when arms control is broadly defined to include any military-related measures that seek to reduce the likelihood of war between adversaries. For example, in their seminal book *Strategy and Arms Control*, Schelling and Halperin (1985: 3) defined arms control as "all the forms of military cooperation between potential enemies in the interest of reducing the likelihood of war, its scope and violence if it occurs, and the political and economic costs of being prepared for it". Gallagher (1998: 1–2) also employed a broader perspective, defining arms control to encompass "any type of cooperative measure meant to reduce the costs and risks associated with the acquisition, threat, and use of military force". This includes "legally binding restrictions on particular weapons, reciprocal unilateral restraints on destabilizing capabilities or practices, and bilateral or multilateral efforts to address the root causes of insecurity". According to these definitions, the difference between arms control and CBMs is blurred, as many CBMs could be then viewed as arms control arrangements.

This has led some scholars to view CBMs as one basic component of arms control, and more precisely as the operational side of arms control. They distinguish accordingly between two main types of arms control: (i) structural arms control, which regulates the types and amounts of weapons systems that states maintain, and (ii) operational arms control, which regulates the operations of military forces and military behaviour such as the level and scope of troop deployment and the rules of military exercises, and includes CBMs (Macintosh 1987: 16; Kemp 1991: 152, 179; Gray 1992: 9; McFate et al. 1994: 15–16; Krause 1997: 185–192; Tanner 2000: 190–191; Heller 2000: 160–162; Schofield 2000: 762). This book will subscribe to a broader arms control definition that contains both structural and operational dimensions. This definition has the merit of placing the question of arms control in its wider Middle Eastern strategic framework.

2.1 Theoretical Approaches

Members of the arms control community are deeply divided over basic questions such as the impact of arms control on the structure of inter-state security relations, the values that arms control should promote, and the means by which it can promote them. Conflicting assumptions about international politics are embedded in arguments over arms control even though they are rarely explicitly identified, contrasted, or tested against each other. In fact, one can conceptualize recent debates over the question of arms control in terms of four main approaches roughly defined by beliefs about the role of arms control in enhancing security and the means through which it can accomplish its objectives. These approaches could be categorized into two groups: academic versus policy-oriented approaches. The criterion of distinction is not the identity of the advocates, but rather the areas of emphasis. Scholars have advocated both academic and policy-oriented approaches. Accordingly, the distinction between the two categories is based on the main thrust of the argument, whether it is rooted in academic traditions and/or theories of international politics or it is more or less oriented towards the advocacy of a certain policy. Indeed, existing approaches are all advocated by academics, but their area of emphasis varies from one approach to the other. Further, some scholars advocate more than one approach at the academic and the policy-oriented levels. Sometimes, academics turn into advocates of certain policies or attempt to project their theories into actions, which explains the recurrence of certain names in more than one approach.

At the academic level, one can identify three major approaches. These include (i) the *Traditional approach*, which views arms control as an instrument of managing and stabilizing security relations between states (ii) the *Transformationalist approach,* which views arms control as an instrument of changing political perceptions as well as security relations between states, and (iii) the *Contextualist approach*, which relates the arms control process to the parameters of the strategic environment. At the policy-oriented level, one could refer to the *Relevancy approach*, which addresses the relevancy/applicability of arms control to conflict-ridden regions. An in-depth analysis of these approaches is in order.

2.1.1 The Traditional Approach

This approach dominated the discourse of arms control for most of the Cold War era. Although there is no realist account of arms control in the literature, this approach is largely informed by the premises of contingent realism, an important variant of structural realism that seeks to explain why states might engage in cooperative arrangements.

At the far end of the realist spectrum lies structural realism which envisions no room for cooperation between states in an anarchic, self-help international system. According to this view, international anarchy forces states to worry about relative military power and to reject negotiation and the upholding of strategically significant limits on their military capabilities. Advocates of structural realism are therefore sceptical about the utility of arms control, as they view the balance of power as a self-sufficient and self-perpetuating system of international security which is to be preferred to arms control (Gray 1992).

In a partial deviation from structural realism, contingent realism does not take a purely zero-sum view of international politics and maintains that cooperation could be the preferred option for states under certain conditions. In his account of contingent realism, Glaser (1995) used the assumptions of structural realism to argue that rational state-actors in a self-help system would opt, under certain conditions, for cooperative policies. He assumed that anarchy is the main characteristic of the international system, and that sovereign states remain the main actors in international politics, have a mixture of conflicting and common interests, and prefer to depend primarily on self-help security strategies. Under these conditions, however, states might choose to cooperate in limited areas where they have common interests. This is especially the case when states face a situation of uncertainty about each other's military intentions. Although uncertainty generates insecurity, and therefore compels states, according to structural realism, to compete, Glaser argued that uncertainty could also create reasons for states to cooperate. Under conditions of uncertainty, states might prefer cooperation over competition if cooperation "reduces the adversary's insecurity by reducing the military threat it faces", or if it "can reduce the adversary's uncertainty, convincing it that the first state is motivated more by insecurity than by greed".

The policy implication of this argument is that states might agree to impose certain limitations on their armament and exchange information about military forces as a means of reducing insecurity and removing uncertainty, and accordingly stabilizing the balance of power system. Although advocates of this approach do not believe that a comprehensive cooperative security system is possible under a self-help system, they favour partial arms control measures within the framework of ad hoc agreements to address security problems that might lead adversaries into an arms race or war that neither side desires.

It is in this context that contingent realists establish a role for arms control in international politics. According to this account, the essential postulate of arms control is the recognition of the possibility of cooperation even between potential adversaries with respect to their military establishments. This cooperation, if it takes place, does not come at the expense of the military security of either actors, as sceptics of arms control would argue. Rather, arms control, if properly conceived, enhances security, especially in the nuclear age where the conception of a good military policy has changed from the purpose of winning wars to that of how to avert them. In this respect, the practice of arms control rests on a theory of crisis stability, which refers to the absence of military incentives to pre-empt under crisis conditions. This is achieved through two fundamental means.

The first means is by strengthening retaliatory, second-strike capabilities that would reduce the possibility of a considered, deliberate military attack. This is based on the assumption that the danger of war would be reduced if both sides were immune to surprise attack. As explained by Jervis (1993: 245), "A first-strike advantage, coupled with the belief that war is very likely, would make it rational for a state to attack even if it was peaceful because the alternative to attacking would be seen as being attacked. This means that the prospects of war will be reduced if both sides posses invulnerable second-strike capabilities." A central goal of arms control in a bilateral or multilateral context is therefore to minimize first-strike advantages by encouraging the building of secure second-strike capabilities on both sides. The second means is by clarifying the military intentions of states in periods of peace and crisis by providing assurances regarding the purpose and character of military activities, in turn inhibiting opportunities for surprise attack, political intimidation, or the outbreak of war by misperception.

In this context, arms control is viewed as a 'security management approach' that aims at stabilizing inter-state security relations across specific parameters. According to this view, although the structure of inter-state relations is still influenced by deterrence and balance of power considerations, participating states would decide, for reasons of their own self-interest, to implement arms control measures in order to stabilize and to lower the cost of a military balance of power. Hence, arms control is to strengthen an existing balance of power, and is not a step toward replacing that balance of power with some different political structure. In other words, cooperation among states in the field of arms control is confined to the stabilization, and possibly the improvement (but not the change or elimination) of a deterrent relationship (Bowker and Williams 1985: 609–610; Rittberger et al. 1990: 70; Schofield 2000: 775; Brauch 2000: 31, 45).

This has led many Traditionalists to link arms control to the logic of security cooperation, rather than security regimes. Whereas cooperation in the context of security regimes takes the form of an institutionalized structure with formal commitments to the implementation of a set of agreed rules and principles, the concept of security cooperation is more flexible in nature since it could include both formal (institutionalized methods of cooperation) and informal cooperation where states might choose to cooperate on specific tactical issues for mutual benefit without committing themselves to the development of formal structures of cooperation.

It is due to this formal/informal nature that cooperation in the field of security becomes possible between both allies and adversaries, with formal structures of cooperation being more common between allies, while informal cooperation is the preferred option between adversaries. Accordingly, the fact that arms control is employed between states to avoid crisis escalation and/or war by misperception in an adversarial security environment, while not affecting the traditional security paradigm, makes arms control more compatible with the concept of security cooperation rather than security regimes.

2.1.2 The Transformationalist Approach

Despite its dominance in the field over the past few decades, the Traditional approach was challenged by a number of scholars whose thinking was largely influenced by the end of the Cold War and the beginning of a new strategic East–West relationship by the early 1990s. The peaceful end of the Cold War, along with the conclusion of a series of American–Soviet arms control agreements over the 1970 and 1980s, brought with it a new understanding of the functions of arms control. According to this, arms control was viewed as a set of instruments that not only maintained the stability of East–West relations during the Cold War but also paved the way for the transformation of these relations (Krause 1997: 191). The impact of arms control, it is argued, was not limited to the achievement of narrow strategic objectives such as avoiding surprise attack and inadvertent escalation, but rather extended to include the transformation of East–West security relations towards cooperation and mutual confidence (Attina 2001: 29–30).

This view is largely influenced by the school of liberal institutionalism; a variant of the liberal paradigm of international relations. This school emphasizes the centrality of international regimes for ensuring free interaction among actors and global application of norms. Some of the advocates of this model have introduced the concept of 'cooperative security approach', according to which states would seek to cooperate in the maintenance of security. This is done through working out a set of collectively binding agreements and normative rules with the purpose of regulating their behaviour within specific-issue areas. In other words, individual states would maintain their security through institutionalized cooperation for the collective handling of problems or conflicts (Haggard and Simmons 1987: 495; Attina 2001: 20–28). Liberal institutionalism has, therefore, moved international security from a system based on deterrence to a new system based on reassurance and transparency.

In this context, proponents of the Transformationalist approach reject the traditional understanding of arms control as limited and incomplete. Instead, they view arms control as one avenue of cooperation among states that leads to the removal of the security dilemma. This is achieved by linking the development and implementation of arms control to a process of transformation in the way participating states think about security relations. Accordingly, the function of arms control goes beyond the stabilization of military balances to include a process of change in the political perceptions of states as well as in the content of their security relations (Macintosh 1996: 31).

The Transformationalist approach also distinguishes between the activity (process) and the end product (measures) of arms control negotiations. According to this view, arms control is conceptualized as a distinctive activity entailing a comprehensive process of exploring, negotiating, and implementing a set of military-oriented measures between states. This activity goes beyond the operationalization of an arms control agreement, and thus should not be confused with what the measures themselves do. It includes three main processes that reinforce each other in triggering the desired political and security transformation. These are: (i) the development/exploration phase, in which the question of arms control is explored and debated informally between academics and security experts on both sides. This debate can take place within the framework of academic conferences, workshops and joint research projects, meetings between journalists or civil society representatives, and military-to-military meetings; (ii) the negotiation phase, in which informal forums and non-governmental discussions are replaced by formal negotiations on specific arms control proposals; and (iii) the implementation phase, in which the participating states implement a negotiated arms control agreement. The argument here is that even though contending states might not accept arms control measures that do more than codify existing defence plans, the countless hours spent on exploration, negotiation and exchange of ideas help both sides recognize their interdependence, understand each other's security concerns, and realize the need for a fundamentally more cooperative approach to security. This multidimensional process of arms control activity is expected to pave the ground for the transformation of security relations from adversarial relations dominated by suspicion and mistrust to a moderated and more cooperative pattern of relations (Macintosh 1996: 36; Steinberg 2004: 263–267; Desjardins 1996: 18).

In this respect, arms control is conceptualized by the proponents of the Transformationalist approach as a change-oriented measure that could trigger a change in the strategic environment between contending parties. The occurrence of this change, however, is not automatic. Rather, it is linked to the presence of a number of supporting conditions, including (i) an overall dissatisfaction with the status quo, its costs, and its security implications among policymakers on both sides; (ii) the presence of a network of experts (epistemic community) that cuts across official and academic lines and is willing to explore and promote the ideas of arms control within the contending states. The significance of the epistemic community is that it can provide policymakers who are dissatisfied with the status quo with new options, including arms control, for dealing with the security problem; and (iii) the occurrence of a positive shift in existing political thinking from an excessive reliance on traditional security schemes toward the adoption of more cooperative security ideas. This entails the rise of a new generation of mid-level, flexible policymakers who are more willing to embrace new, cooperative ways of maintaining security (Macintosh 1996: 37–38).

Advocates of the Transformationalist approach also view arms control from an international regime perspective. Conceptualized as an instrument that would create new standards of behaviour (norms) for the rules of political engagement and the management of security relations, arms control is thus linked to the establishment of international security regimes. The linkage is established not in the sense of having arms control creating a legal framework comparable to international law, but in the sense of establishing a quasi-law in the form of mutually agreed rules of the game or code of conduct which allow for building confidence and the creation of more cooperative patterns of maintaining security among the participating states. In this respect, the concept of arms control has become an integral part of the model of international regimes and the idea of an arms control regime has been introduced.

2.1.3 The Contextualist Approach

The Contextualist approach addresses the question of arms control from a different perspective. In the Traditionalist–Transformationalist debate, scholars diverge on the impact of arms control on inter-state security relations. Whereas the Traditionalists contend that arms control is to stabilize these relations along the parameters of deterrence and balance of power, the Transformationalists argue that arms control can change/transform such relations from adversarial to more cooperative patterns. In this context, arms control is treated as a prelude to other political and military developments. The policy implication of these propositions is that arms control, once applied in a bilateral or multilateral context, will affect existing patterns of security relations between states, either by stabilizing or transforming them. In the Contextualist approach, however, arms control is viewed as a reflection, rather than a cause, of the regional or global setting in which it is introduced.

Proponents of the Contextualist approach question some of the propositions underlying the concept of arms control; chiefly among these is the proposition that armaments contribute significantly to the outbreak of war. According to this proposition, arms races are viewed as an important link in the process of conflict escalation, which is likely to result in the outbreak of war between opposed powers. The policy implication here is that arms races should be brought under control and eventually reversed if war is to be avoided. However, for the advocates of the Contextualist approach, this conclusion is not without limitations. This is because whereas armaments are among the conditions that enable wars to take place, they do not necessarily produce war, or provide in themselves a means of distinguishing the conditions of war from the conditions of peace. In fact, not all arms races have been followed by wars. Instead, some races have come to an end, as was the case with the American–Soviet arms race during the Cold War. Even in cases of war, the fact that some wars were preceded by arms races could suggest a correlation between the two variables, but it does not guarantee the existence of a causal relationship. In addition, historical experience shows that the application of arms control, particularly CBMs, in bilateral or multilateral settings has not always been met with success. In some cases, arms control has failed to achieve its desired objectives, either those related to conflict management and crisis stability or those related to the changing of political-security perceptions.

Further, arms races may be outcomes rather than causes of conflicts. The conflict of interests and the perception of grievances on the part of some of the contending parties may motivate them to arm in order to rectify perceived injustices. In this case, an emphasis on the arms race without reference to the underlying causes of the race might lead to misleading conclusions. What follows is that any attempt to establish an arms control regime must take into account the wider regional context in which such a regime is to be installed (Tuchman 1984: 13–14). Arms control does not automatically improve or eliminate antagonistic security structures that are determined by opposing interests or even hostility. Rather, this impact is largely determined by the occurrence of certain developments in the regional political and security environment, which, in return, act as the enabling conditions for arms control. This makes the Traditionalist–Transformationalist debate over the impact of arms control on security largely irrelevant to the analysis as it does not address the conditions under which arms control becomes applicable within a conflict situation.

In fact, advocates of the Contextualist approach have come up with different assessments of the enabling conditions for arms control. In his classical work, Bull (1965: 7–10, 65–79) identified three important conditions that make regional settings conducive for the establishment of arms control regimes. These include (i) that the powers concerned want a system of arms control; (ii) a measure of political détente among them sufficient to allow for a system of arms control; and (iii) a mutual interest in the military situation that the arms control process will legitimize. Richter (1994: 72) subscribed to a similar view. He contended that arms control, especially CBMs, are most likely to have little or no impact on eliminating the causes of tensions and improving security relations between states if they are not introduced within the proper political context. Richter emphasized the element of détente as an essential precondition for the successful application of CBMs, arguing that CBMs cannot in themselves eliminate structures of antagonistic security relations. Accordingly, CBMs have the potential to create cooperative security structures only when they are introduced within the context of a comprehensive process aimed at gradually eliminating political differences.

Blacker and Duffy (1984) also contended that arms control must be preceded by the creation of mutually acceptable military conditions, since no country will accept or comply with treaties unless they are in its own interests. According to their analysis, progress in arms control can only be achieved with the existence of a balance of power between the contending parties. Blacker and Duffy supported their argument with reference to the historical experience of arms control. In the Cold War arms control process, the launching of the first arms control talks between the two superpowers in 1969 and the conclusion of the SALT I agreement in 1972 and other subsequent agreements became possible after the reaching of a strategic balance of power between NATO and the Warsaw Pact. For most of the 1960s, the Soviet Union was suspicious of entering into any arms control talks with the United States because of Soviet relative nuclear inferiority and the fear that arms control would perpetuate such inferiority. It is only when the Soviets were in a process of achieving nuclear strategic parity with the United States by 1969 that they expressed willingness to negotiate arms control agreements. With the reaching of the balance of terror, the question of arms control gained momentum in the policy and military circles of both NATO and Warsaw Pact as a means of controlling the nuclear arms race and of preventing a deadly nuclear confrontation. The fear of mutual destruction convinced the United States and the Soviet Union that a limitation on their nuclear weapons was in their best interest.

A similar pattern could also be observed in other arms control agreements before World War II. These included (i) the 1817 American–British Rush–Bagot agreement on the reduction of their naval forces on the Great Lakes. The agreement, described by Blacker and Duffy as "the most successful disarmament effort of the nineteenth century", was concluded and sustained within a framework of naval parity between the United States and Britain; and (ii) the 1922 Washington Naval Treaty between Britain, the United States, Japan, France and Italy on halting the construction of new capital ships and aircraft carriers over a ten-year period, and its extension in the London Naval Treaty of 1930 between Britain, the United States and Japan. The Washington treaty created naval parity between Britain and the United States by limiting the total tonnage of their capital ships to 500,000 tons for each, while it put Japan, France and Italy at a disadvantage by limiting their capital ships' total tonnage to 300,000, 175,000, and 175,000 respectively. This uneven distribution of naval capabilities led eventually to the collapse of the Washington and London naval systems. Whereas France and Italy expressed reservations about the distribution ratio and blocked efforts in 1927 to extend the agreement to cover cruisers and auxiliary vessels, Japan became more unsatisfied with the agreements, arguing that they would perpetuate its naval inferiority vis-à-vis the United States and Britain. In 1934, Japan formally withdrew from the Washington and London agreements, and in 1935 the two agreements practically collapsed when the London Naval Conference concluded a new agreement that virtually reversed the limitations stipulated by the 1922 and 1930 agreements.

Bromley and Perdomo (2005) viewed the enabling conditions from a different perspective, arguing that arms control is largely dependent on two interrelated factors. The first is the type of regime. According to their analysis, stable democratic regimes are important to ensure accountability and commitment to arms control agreements, whereas in weak democracies or non-democratic regimes, arms control obligations are less likely to be met in a consistent and coherent way. The second is the existence of a shared political culture among the states involved. States with similar political cultures are more inclined to respect their commitments under cooperative frameworks, including arms control regimes. In areas with antagonistic political cultures, there is a high risk of divergence, and this could limit the potential impact of arms control. In the absence of these conditions, the authors argue, there is a tendency to question states' commitment to the implementation of arms control agreements, as well as a higher probability of cheating.

2.1.4 The Relevancy Approach

The Relevancy approach addresses the question of the applicability of operational arms control (CBMs) in other conflict-ridden regions outside Europe. Under this approach, one can articulate two general trends. The first trend, which is more dominant in the literature, suggests that the European experience of CBMs could be applied in other regions. This suggestion has some methodological relevance. One of the major assumptions of social science research is that human behaviour is patterned. Human experiences occur in the form of patterns and, therefore, they are generalizable. In this context, one of the objectives of social science is to discover these patterns and use them to explain future human behaviour. With this basic assumption, some analysts have introduced the notion of trans-regional learning, which refers to the possibility of drawing inferences from one region and applying them to others, with arguments that the experience of one region could be used to understand the dynamics of other regions. Accordingly, they have advocated the applicability of the European experience of CBMs to other non-European regions, emphasizing that European-style CBMs can stimulate problem-solving approaches in other geographical settings. According to the advocates of this trend, as CBMs had succeeded in stabilizing East–West relations and in setting the ground for the peaceful end of the Cold War, these measures could also achieve similar successes in other conflict-ridden regions.

This view has been advocated by a number of scholars, both from the Traditionalist and Transformationalist camps. For example, Brauch (2000a: 333–334), a strong advocate of the Traditionalist approach, maintained that CBMs are important tools for crisis stability in different regions of the world. He paid special attention to the southern Mediterranean region as one of the most unstable and conflict-ridden regions in the world, arguing that the region is in need for CBMs in order to defuse tension in times of crisis. Kemp (1991: 170–181) from the Transformationalist camp emphasized the significance of CBMs in the Middle East region. He advocated the application of CBMs and structural arms control measures between Israel and her Arab neighbours in order to trigger a process of change towards the resolution of existing conflicts and the prevention of future ones. Ahmar (2001: 43) also contended that CBMs might be a European creation, but the philosophy and content of CBMs should not be restricted to the European context. He explained that "CBMs do not have an inclination towards any specific community or group, but possess the ability to deal with other crises and conflicts in different parts of the world". He added that "The concept of CBMs originated in the West but its application is universal in nature".

The second trend rejects the notion that CBMs are universal conflict-management or conflict-resolution tools. This trend is largely informed by the assumptions of the Contextualist approach, and also has some methodological relevance. Despite the significance of trans-regional learning, social scientists have warned against "the use of geography on the assumption that the various areas are the same" (Simon 1978: 179–181). Geographical differences must be taken into account if one is to make valid inferences and generalizations. In other words, geographical regions must be similar in the most crucial dimensions in order to be able to compare these regions or draw inferences from one region and apply them to others. Further, one of the pillars of social science research is the concept of antecedent variables, which refers to "the kinds of conditions under which the original relationship was at least and most likely to occur, and the kinds of processes that were involved in the operation of the original relationship" (Selltiz et al. 1976: 45). Accordingly, social scientists argue that in validating a relationship, one must specify the conditions or contingencies necessary for the occurrence of the relationship. Such conditions (contexts) include three major elements: interest and concern, time and place, and background characteristics. People differ in their concerns and interests, which in turn affects their attitudes and behaviour patterns. In addition, a relationship between two variables can vary according to the time and place in which it is studied. Similarly, associations are likely to differ for persons or groups that do not share the same characteristics (Nachmias and Nachmias 1992: 410–412).

In this context, proponents of the second trend argue that CBMs are a European phenomenon that emerged within a particular political and historical context in order to serve European security requirements. Conflicts in other regions of the world, it is argued, have different dynamics from those of the East–West conflict during the Cold War, and, as a result, CBMs could not address the security problems of other conflict-ridden regions. This led Richter (1994: 73) to warn that any attempt to employ CBMs in other regional (non-European) settings must be made with caution. He built his argument in light of two important factors. The first is the profound differences between the political structure of Europe during the Cold War and that of most developing regions, with the latter being usually characterized by cross-cutting conflict dyads, unstable patterns of conflict, incomplete state formation, and weak sub-regional integration. These characteristics would make it very difficult to implement CBMs since the "transaction costs for the establishment of an all-regional conference process would likely be very high, whereas the capacity of individual states to raise the funds to pay for these costs is low". The second is the political and historical preconditions of CBMs, which are generally lacking in the developing world. Whereas Europe witnessed a high degree of mutual strategic deterrence, and a process of political normalization that led to the renouncement of violent change of existing borders, prior to the rise and implementation of CBMs, most, if not all, other regional settings have lacked such strategic developments. Richter went further to warn against the potential negative outcomes of projecting CBMs to other regions:

> Analysts should recognize that the effort to implement CBMs in other regions can become counter-productive, effectively reducing the level of confidence and trust in the interactions of various regional actors, rather than improving it. This is because a poorly executed and conceived attempt to develop a confidence-building regime can precipitate the very types of suspicion, mistrust, and misperception that confidence-building is supposed to correct (Richter 1994: 62).

2.2 Assessment of the Arms Control Approaches

No doubt, these scholarly approaches shed light on important dimensions of arms control and its functions in conflict situations. However, one could raise a number of critical remarks about the arms control literature in terms of its theoretical assumptions, methodological basis, and the domain of analysis. These include the following:

2.2.1 Eurocentric Accounts

The arms control literature depends excessively on the Cold War experience of arms control as the main domain of analysis. Although arms control might find some grounds for exploration and application in a number of developing regions, these efforts have not received enough attention in the literature as they have been too limited in scope and/or they have not succeeded in laying the foundation of a genuine arms control regime. This Eurocentric view is particularly evident in the CBMs literature, which treats the European experience as the main, if not the only, testing ground for drawing theoretical and policy-oriented conclusions. This trend, however, could be criticized on two main grounds.

First, although the concept of CBMs emerged officially within the framework of the European CSCE process in 1975, the idea of CBMs was not unfamiliar to a number of developing regions over the past few decades, and even before the conclusion of the Helsinki process. An obvious example here is the Israeli–Egyptian CBMs which were implemented in the context of the first and second disengagement agreements in 1974 and 1975 respectively, and the Egyptian–Israeli peace treaty in 1979. Under the peace treaty, a comprehensive set of 'CBMs' were applied on both sides of the Egyptian–Israeli borders. This included the establishment of demilitarized zones, hotlines, limitation of forces, early warning stations in designated areas, and monitoring, surveillance and inspections of military formations and troop movements.[3] In fact, the Egyptian–Israeli experience of CBMs represents one of the most successful examples of CBMs outside Europe.

One could also refer to other success stories of CBMs, which include (i) the Russian–Chinese agreement to develop a CBMs regime for their border region as part of a comprehensive agreement signed between the two countries, as well as Kazakhstan, Tajikistan, and the Kyrgyz republic, in 1996, and (ii) the development of a number of legally binding CBMs agreements in Latin America, the most important of which were the 1997 Inter-American Convention Against the Illicit Manufacturing of and Trafficking in Firearms, Ammunition, Explosives, and Other Related Materials, and the 1999 Inter-American Convention on Transparency in

[3] For a detailed review of CBMs within the context of the Egyptian–Israeli Peace Treaty, see the text of the "Egyptian–Israeli Peace Treaty", 1979; available at: http://www.mideastweb.org/egyptisraeltreaty.htm.

Conventional Weapons Acquisitions. Whereas the first agreement focused on enhancing military stability and accountability, the second was directed towards inter-state security and crime prevention.

Second, the fact that CBMs had limited, or even unsuccessful applications in non-European settings does not suggest that these cases are irrelevant to the analysis. Rather, such cases could provide meaningful insights into the analysis of CBMs, and particularly the study of the potential constraints that might hinder the application of CBMs within conflict situations. This is the case, for example, with the Middle East and South Asia. Since the outbreak of the Arab–Israeli and Indo-Pakistani conflicts in the Middle East and South Asia respectively, the two regions had experienced the application of several types of CBMs for the purpose of ending military confrontations and reaching a political solution through third party mediation and direct negotiations. In the Middle East, CBMs began with the end of the first Arab–Israeli military confrontation in 1948 and have continued ever since at different intervals, although they were not labelled as such until the end of the Cold War. These measures were included within formal and informal agreements between the parties to the conflict. With the end of the Cold War, the jargon of CBMs reappeared at a much higher level in the Arab–Israeli discourse as a result of the beginning of the Madrid peace process in 1991 and subsequent negotiations between the parties to the conflict. The resurgence of CBMs was also a result of Euro-American initiatives to transfer the Helsinki process to the Arab–Israeli conflict. In South Asia, several CBMs were also included within formal bilateral agreements between India and Pakistan in order to end the state of war and reach a political solution to the Kashmir problem. Indeed, one could describe CBMs in the Middle East and South Asia as the fastest-growing business witnessed by the two regions over the second half of the twentieth century.

However, CBMs proved to be a failing business in the Middle East and South Asia, both of which still exist today as among the most conflict-prone regions in the world. Except for the Egyptian–Israeli 'CBMs' in 1979, all CBMs introduced in the context of the Arab–Israeli conflict during the Cold War failed to prevent the outbreak of military confrontations between the antagonists. Post-1949 measures were not an impediment to the outbreak of the 1956 Suez war, and post-1957 measures did not block the June 1967 war. These measures prevented specific crises, but did not lead to the resolution of the conflict, did not remove misperceptions, and did not even constrain the ability of some of the parties to go to war. Further, the post-Cold War CBMs initiatives achieved limited success. Whereas Israel endorsed them, the Arabs perceived them as instruments for consolidating Israeli military hegemony and occupation of their territories, and thus were reluctant to accept them. In South Asia, CBMs also proved to be a failing business. Despite decades of institutional CBMs between India and Pakistan, there still exists deep-rooted suspicion, dispute, and insecurity between the two countries. CBMs did not also safeguard against the eruption of crises or the escalation of conflicts. Almost each round of CBMs was followed by a round of military confrontation. Although the two parties have not gone to war against each other since 1971, they have clashed several times over Kashmir in the form of 'small-scale confrontations'.

This failure of CBMs raises a number of questions regarding the role of CBMs in the process of conflict resolution; chief among them are (i) why did CBMs succeed in the European context, but fail in the Middle East and South Asia? and (ii) do CBMs represent a Western phenomenon that serves specific European security requirements and thus are not applicable to other non-European regions, or do they act as universal tools that could be imitated in other conflict-ridden regions but whose application is linked to the presence of a conducive political and strategic context? No doubt, an examination of why CBMs have not been successful in some parts of the world should help us to move beyond European boundaries, and, accordingly, to enrich our theoretical understanding of the concept and its dynamics.

2.2.2 Lack of Well-Developed Theoretical Accounts

The arms control literature is characterized by the lack of theoretical accounts that help explain the dynamics of arms control in conflict situations or within the framework of the conflict resolution process. In fact, one could argue that the field has witnessed limited academic effort addressed at locating the concept of arms control within a larger theoretical framework. The focus, instead, tends to be on the empirical dimension and policy implications of arms control. This makes the literature largely atheoretical as it does not possess a well-established theoretical perspective to structure an understanding of the dynamics of arms control.

For example, the Relevancy approach is essentially atheoretical, as it deals with the viability of arms control for conflict resolution in specific areas. The Traditional approach also does not show *how* and *why* arms control works between states. For example, if the main function of arms control is to stabilize an existing deterrent relationship, does this mean that arms control is an attractive instrument to any two members of a dyad whose interaction is stipulated by a deterrent relationship, or are there certain (intervening) factors that should be present in order to arouse interest in arms control as a stabilizing instrument? Further, if arms control is to achieve crisis stability, remove misperceptions and reduce insecurity between contending states, then why has it little more than minimal impact on political and security perceptions in an unpredictable security environment?

Similarly, the Transformationalist approach has theoretical shortcomings. Although it tries to address many of the pitfalls of the Traditional approach by insisting on asking *why* and *how* arms control is expected to produce a certain impact on inter-state security relations, the Transformationalist approach also leaves many important questions unanswered. For example, in arguing that the impact of arms control is linked to the occurrence of a 'positive shift' in political thinking, the Transformationalist approach does not address the following questions: (i) what are the factors responsible for the occurrence of such a positive shift? (ii) are these factors linked to changes within the domestic environment, such as the change of leadership, or are they linked to changes at the regional level, such as a shift in the balance of power or the settlement of political differences? (iii) how can the contending parties realize that a positive shift in political thinking is taking place? (iv) are there any signs that could help identify this shift? and (v) if so, what are these signs? Indeed, Macintosh (1996: 66), a strong advocate of the Transformationalist approach, recognized many of these shortcomings when he admitted that this approach should not be viewed as constituting a theory of international relations, despite efforts to place it within the broader accounts of international institutions. Instead, he recognized the approach as a "much more modest and limited conceptual creation".

In addition, one could argue that the most serious theoretical problems are to be found in the Contextualist approach. Although it addresses an important dimension of the arms control debate by looking into the enabling conditions for arms control, there is no consensus among the advocates of this approach on defining these conditions. Whereas some scholars view political détente as the most important for an arms control process to start, others contend that it is the balance of power which matters, arguing that the achievement of strategic equilibrium between states leads to a reduction of tensions and the creation of a general atmosphere of political détente, which eventually paves the way for an arms control process to begin. Still a third group of scholars contends that a shared democratic culture is the key requirement for the success of arms control. Also, the Contextualist approach does not address the exact link between the enabling conditions and arms control. Does the application of arms control follow directly after the enabling conditions materialize, or does it also depend on the presence of other variables in either the domestic or regional environments? More importantly, the Contextualist approach does not examine the impact of arms control, once applied, on the structure of political and security relations between the participating states. Does arms control function as a measure oriented towards the status quo and seeking the stabilization of security structures across specific parameters, or does it act as a change-oriented measure whose purpose is to transform existing structures? Does arms control function as a tool for crisis stability, conflict management, or conflict resolution? In other words, if the Contextualists criticize others for not paying attention to the political and strategic context under which arms control should operate, they themselves have fallen into the trap of not addressing the role of arms control between states.

2.2.3 Failure to Recognize Complexity Within Operational Arms Control

The literature treats operational arms control as a unitary concept. Although the literature classifies CBMs into different categories, such as communication, notification, and inspection measures, it fails to acknowledge the differences between these categories regarding their impact on political-security structures as well as the conditions under which they could be introduced. Instead, the classification is made for the mere purpose of distinction, with no attempt made to analyse how and why each category works in a given conflict situation. This is the case, for example, with

(i) communication measures, such as hotlines, and (ii) inspection measures, such as on-site inspection. Although both measures are classified as CBMs, each deals with different security requirements, which means that the preconditions for their application are also different. On the one hand, hotline measures are considered the fastest-growing CBMs in terms of application because they are seen as harmless and risk-free, and thus they are the most attractive to states involved in adversarial relations. On the other hand, inspection measures are considered the least common because they deal with the verification of actual military activities and structures, and thus they are seen as involving 'potential' risks for national security and state sovereignty. Accordingly, the application of verification measures require the existence of a moderate and more cooperative pattern of relations between the participating states in order to be able to overcome any possible rise of threat perceptions.

Indeed, historical experience shows that different types of CBMs have served different purposes under specific political and strategic contexts. In Europe, the progression of CBMs within the framework of the CSCE process was a response to the gradual development of East–West relations to more cooperative patterns towards the late 1970s and 1980s. In 1975, the Helsinki Final Act officially adopted the first generation of CBMs in what became known as *The Document on Confidence-Building Measures and Certain Aspects of Security and Disarmament*. The Document endorsed the application of a number of CBMs among the CSCE members, including (i) prior notification of major military manoeuvres, (ii) prior notification of other military manoeuvres (less than 25,000 troops), and (iii) exchange of observers. The Document stipulated that states conducting military manoeuvres should invite other CSCE states, voluntarily and on a bilateral basis, to send observers to attend the manoeuvres.[4] However, the Helsinki Document occasioned a number of critical remarks, the most important being the non-binding character of CBMs and the absence of verification provisions. This meant that a CSCE member state could conduct military manoeuvres without notifying other members. There was also no mechanism whereby each party could verify the information received from the other. This created the fear that CBMs could be used by the protagonists as tools for deception, provoking false confidence.

These concerns motivated the CSCE countries to come up with the second generation of CBMs in the CSCE Stockholm Conference of 1986 in what became known as the *Conference on Confidence and Security-Building Measures and Disarmament in Europe* (CDE). The Stockholm Document strengthened the provisions of the Helsinki Final Act and extended the scope of CBMs to the area of verification. In Stockholm, the CSCE participants agreed on (i) extending the time for prior notification from twenty-one to forty-two days, and manoeuvres of a minimum of 13,000 men instead of 25,000 were subject to prior announcement, (ii) the exchange of annual calendars for military activities subject to prior notification, and more importantly (iii) the adoption of verification measures, which were extended

[4] See the text of "The Helsinki Final Act", 1975; available at: http://www.osce.org/mc/39501?download=true.

to include on-site inspection.[5] . In fact, the adoption of on-site inspection, which covered inspection of military airports, military establishments, and military bases, represented a major breakthrough in the development of CBMs. The objective was to enable each party to know what the other was doing and why, to know the size of troops, firepower, troops structure, weaponry systems, and training procedures, and thus to guarantee that seeing was indeed tantamount to believing.

This progression into more significant, verifiable, and binding CBMs was not a seamless development in which NATO and Warsaw Pact moved automatically from Helsinki to Stockholm. Rather, it was the result of the second era of détente. In 1985, Mikhail Gorbachev came to power in Moscow with a new strategic vision for relations with the West. Gorbachev realized there were two important realities about the Soviet Union: (i) that the worsening of economic conditions in the Soviet Union during the Brezhnev years had undermined its capacity to project power and influence at the global level, and (ii) that the increasing technological gap between the Soviet Union and the United States in favour of the latter meant that the Soviet Union would be losing ground in the continuation of the arms race. In this context, the new Soviet leadership developed a new strategic outlook which evolved around the notion of the 'balance of interests' rather than the balance of power. In November 1985, an American–Soviet summit was held. The summit was the first between the two superpowers since 1979. Another summit was held in October 1986 in which the two sides discussed proposals for arms control agreements. It was in the context of the second détente that the CSCE countries were able to implement harsher CBMs which involved potential risks for national security and state sovereignty. It was only then that Gorbachev gave his signal for approval that the Soviet Union would accept the introduction of verification measures into the Stockholm Document.

Similarly, the third and fourth generations of CBMs were an outcome of a set of global transformations that took place in the late 1980s and early 1990s. The American–Soviet arms control negotiations which had started in 1986 resulted in the signing of the *Intermediate-range Nuclear Forces* (INF) Treaty in December 1987. The INF Treaty was the first arms control agreement in a decade. The Treaty stipulated the elimination of all ground-based intermediate and short-range nuclear ballistic missiles from the European theatre. It also established a verification regime based on cooperative monitoring. Four years after the conclusion of the INF Treaty, the superpowers concluded the *Strategic Arms Reduction Treaty* (START I) in July 1991, which was considered the first treaty to actually reduce the size of strategic nuclear arsenals.[6] In addition, the year 1991 witnessed the collapse of the Soviet Union and the end of the Cold War. The end of the Soviet

---

5 See text of "The Stockholm Document", 1986; available at:http://www.osce.org/fsc/41238

6 Under START I, the two superpowers were obliged to cut the number of their strategic nuclear warheads to no more than 6,000 each, which could be deployed on no more than 1,600 strategic missiles and heavy bombers. The treaty also banned the production, testing and deployment of new or modified ICBMs and SLBMs with more than ten warheads, and provided for intrusive verification procedures.

empire brought the nuclear confrontation between the East and West to an end. It also freed East European countries from Soviet domination, and enabled them to pursue a pro-Western policy. It was in the context of these transformations in the global strategic environment that the CSCE members were able to develop more advanced generations of CBMs.

The history of CBMs between Egypt and Israel also represents another case in point. In 1957, following the Suez crisis, Egypt unilaterally introduced a number of security and confidence-building arrangements in Sinai in exchange for a full Israeli withdrawal from Sinai. First, Egypt permitted the deployment of a UN peacekeeping force in Sinai along the borders with Israel and at checkpoints on the Straits of Tiran. The UN force was mandated to carry out patrols, man sensitive border positions, and prevent infiltration across the borders. It also provided assurances against violations. Second, Egypt agreed to carry out demilitarization measures by limiting its troops in Sinai to two military divisions. These CBMs were modest in nature compared with other types of CBMs. Indeed, there was no potential for higher levels of CBMs due to the highly antagonistic pattern of Egyptian-Isreali relations at that time.

Twenty years later, however, the two countries were able to implement higher levels of CBMs. In 1977, relations between Egypt and Israel witnessed a major breakthrough due to the political initiative of the then Egyptian President Anwar El-Sadat, who entered into direct peace negotiations with Israel. The Sadat Initiative, as it came to be widely known, was unexpected and even shocking. For the first time, an Arab political leader was negotiating directly with Israel and visiting that country even though the territorial issues were not resolved. The Initiative changed the strategic environment of the Egyptian–Israeli conflict as it created a momentum toward the resolution of the territorial issues. In this context, Egypt and Israel entered into intensive negotiations under the auspices of the United States, which resulted in the conclusion of the Camp David agreement in 1978 and the Egyptian–Israeli peace treaty in 1979. The peace treaty provided for the resolution of Egyptian–Israeli territorial issues and the normalization of bilateral relations. It also introduced an extensive set of bilateral, verifiable CBMs such as hotlines, demilitarized zones, limitation of forces, early warning stations in designated areas, monitoring, surveillance, and on-site inspections.

2.2.4 Reliance on Poorly-Developed Assumptions

Last but not least, the arms control literature relies on poorly developed assumptions about the nature and operation of arms control. It fails to establish the empirical link between arms control and the structure of inter-state security relations. It rather treats the impact of arms control on security as assumed rather than proven. In the Traditional approach, for example, one of the underlying assumptions about arms control is that it will reduce misperceptions and clarify intentions between states, and thus improve their security relations. This is presumed to take place because the exchange of information about military capabilities and activities between states will lead to greater transparency, which in turn will help reduce suspicions, build confidence, and bring some stabilization or improvement in the structure of security relations.

This assumption, however, is highly questionable on a number of grounds. First, the fact that arms control is treated within the Traditional approach as a mechanism oriented towards the status quo makes it unlikely that arms control will reduce the sources of security risks between states. This is because arms control does not deal with the core issues of contention, basically the main source of conflict. In this respect, arms control might be used to avoid dealing with the actual security problems or, in some cases, to replace the implementation of other measures that could be more effective in addressing the main sources of conflict. More importantly, there is no well-established evidence that a positive relationship exists between information and confidence. The acquisition of more information about the military forces of the adversary is not a guarantee for improving security relations. Indeed, having more information about the adversary through arms control agreements could in some instances feed suspicions rather than resolve them. This is particularly the case when arms control could not provide 'perfect' information in a fundamentally suspicious environment. According to Desjardins (1996: 62),

> Data not confirming information acquired by other means or suggesting a more serious threat than previously believed would not increase confidence, but is more likely to create mistrust and suspicion. Information provided voluntarily, no matter what its quality or accuracy, may not be necessarily believed. In fact, it may very well only bring more questions, apprehensions, and misgivings about what is not known. Not all forms of transparency will necessarily be useful to build confidence. Half-truths about the real purpose of some weapons acquisition are likely to reinforce patterns of suspicions and mistrust, especially if such semi-transparency is under the cover of an agreement designed to reassure others of peaceful intent.

Although the Transformationalist approach tried to look deeper into the way arms control works between states, it also fell into the trap of presumed, rather than scientifically proven, connections. In the Transformationalist view, the argument that an arms control process will trigger a positive change in the security environment is based on a number of interrelated assumptions on how this process is intended to work. Perhaps the most important of these assumptions are (i) that two or more contending parties will have a mutual interest in launching an arms control process; (ii) that the parties will agree on the precise content of an arms control dialogue; (iii) that formal negotiations will result in an arms control agreement; (iv) that agreement will be translated into effective implementation; and (v) that successful implementation will lead to the transformation of political and security perceptions in a positive direction. This set of assumptions is supported by no scientific evidence. There is no assurance that the launching of an arms control process will ultimately lead to the implementation of meaningful arms control agreements, and accordingly the occurrence of positive shifts in attitudes. In fact, there are cases in which the arms control process has failed to move beyond the initial stages of exploration and discussions for a variety of reasons, such as the case of arms control negotiations between the Arabs and Israel in the context of the ACRS working group. In other cases, the launching of an arms control process might in itself lead to negative results if things did not go well during the three main phases (exploration, negotiation, and implementation) of the process, especially towards the last phase where discussions have to lead to the conclusion and implementation of a negotiated arms control agreement. If the contending parties failed to implement the agreement for any reason, this could destroy any sort of confidence built earlier during the exploration and negotiation phases.

The main conclusion to be drawn from this review is that the literature on arms control cannot locate the exact role of arms control within the framework of the processes of conflict prevention or conflict resolution. In fact, the literature is in dire need of the development of a theory which clearly identifies the dependent variable, and places arms control within the overall context of the conflict resolution process. The Traditional/Transformationalist dichotomy reflects the confusion in the field over what arms control actually implies and involves. Does arms control refer to specific measures designed to (i) stabilize security relations and regulate military forces, or (ii) change the content of the strategic environment towards more cooperative security structures? Arms control theorists are not sure about which dependent variable is linked to arms control. Is it crisis stability, conflict resolution, or conflict prevention? The exact role of arms control will differ depending on that variable. Arms control may play a role in crisis stability and conflict prevention, but not in conflict resolution. This confusion has even become worse with the rise of the Contextualists, who have shifted the focus of analysis from an examination of the impact of arms control on the security environment to an examination of the conditions that create an interest in arms control. Within the Contextualist approach, there is no agreement among scholars about the preconditions (the intervening variables) that pave the way for the application of arms control. Is it political détente, balance of power, shared democratic culture, or all of these? Further, there is no agreement on the exact link between these preconditions and the actual implementation of arms control. Do these preconditions create the political will to apply arms control, or do they mainly serve as supporting conditions to a process whose initiation and implementation are linked to other variables?

This state of theoretical confusion is largely reflected in the way arms control has been approached in the Middle East. Since the rise of the arms control agenda in the Middle East in the early 1990s, concerned global and regional actors have developed divergent views about the utility of arms control in the context of the Arab–Israeli conflict. Although these views have been largely motivated by political considerations, their advocates have made frequent references to different theoretical approaches in an attempt to support their argument. Whereas global actors have viewed arms control in the Middle East from a Transformationalist perspective, thereby advocating the application of arms control as an essential component of a fundamentally more cooperative approach to conflict resolution and regional security, the Arabs have been quite cautious about arms control, and viewed it from either (i) a Traditional perspective as an instrument for managing political relations and military balances, and thus not as a substitute for conflict resolution, or (ii) a Contextualist perspective as a code of conduct whose application must be preceded by the occurrence of certain transformations in the strategic environment between the contending parties. This calls for a review of global approaches to arms control in the Middle East, the context in which they were articulated, and the reactions they triggered from major regional, particularly Arab, actors.

References

Ahmar, M. (2001). Rationalizing the concept of confidence-building measures. In M. Ahmar (Ed.), *The challenge of confidence-building in South Asia*. Karachi: Paramount Books Ltd.

Attina, F., & Attina, F. (2001). Partnership and security: Some theoretical and empirical reasons for positive developments in the euro-mediterranean area. In F. Attina & S. Stavridis (Eds.), *The Barcelona process and euro-mediterranean issues from Stuttgart to Marseilles*. Milano: Dott. A. Giuffre Editors.

Blacker, C. D., & Duffy, G. (1984). *International arms control: Issues and agreements*. California: Stanford University Press.

Booth, K. (1987). Disarmament and arms control. In J. Baylis, K. Booth, J. Garnett, & P. Williams (Eds.), *Contemporary strategy I.: Theories and concepts*. Teaneck, NJ: Holmes & Meier Publishers.

Bowker, M., & Williams, P. (1985). Helsinki and west European security. *International Affairs*, *61*,4 (Autumn).

Brauch, H. G. (2000). From confidence to partnership building measures in Europe and the mediterranean: Conceptual and political efforts revisited. In H. G. Brauch, A. Marquina, A. Biad, & P. Liotta (Eds.), *Euro-mediterranean partnership for the 21st century*. London: Macmillan.

Brauch, H. G. (2000a). Beyond Stuttgart: Prospects for confidence and partnership building measures in euro-mediterranean relations for the 21st century. In H. G. Brauch, A. Marquina, A. Biad, & P. Liotta (Eds.), *Euro-mediterranean partnership for the 21st century*. London: Macmillan.

Bromley, M., & Perdomo, C. (2005). CBMs in Latin America and the effect of arms acquisitions by Venezuela. Working Paper (WP) 41/2005, Real Instituto Elcano, 22 September.

Bull, H. (1965). *The control of the arms race: Disarmament and arms control in the missile age*. New York: Fredeirck A. Praeger Publishers.

Desjardins, M.-F. (1996). *Rethinking confidence-building measures: Obstacles to agreement and the risks of overselling the process*. Oxford: Oxford University Press.

Gallagher, N. W. (1998). Bridging the gaps on arms control. In N. Gallagher (Ed.), *Arms control: New approaches to theory and policy*. New York: Frank Cass.

Glaser, C. L. (1995). Realists as optimists: Cooperation and self-help. *International Security*, *19*, 3.

Gray, C. (1992). *House of cards: Why arms control must fail*. Ithaca: Cornell University Press.

Griffiths, M., & O'Callaghan, T. (2002). *International relations: The key concepts*. London: Routledge.

Gunduz, A. (1994). *Security & Human Rights in Europe: The CECE process*. Istanbul: University of Marmara.

Haggard, S, & Simmons, B. A. (1987). Theories of international regimes. *International Organization*, *41*, 3 (Summer).

Heller, M. A. (2000). Weapons of mass destruction and Euro-Mediterranean policies of arms control: An Israeli perspective. In A. Vasconcelos & G. Joffe (Eds.), *The Barcelona process: Building a Euro-Mediterranean regional community*. London: Frank Cass.

Holst, J., & Melander, K. (1977, July/August). European security and confidence-building measures. *Survival*.

Holst, J. (1987). A CBM regime in Europe: 19 interrelated propositions. In R. B. Byers, F. S. Larrabee, & A. Lynch (Eds.), *Confidence-building measures and international security*. New York: Institute for East–West Security Studies.

Jervis, R. (1993). Arms control, stability, and causes of war. *Political Science Quarterly, 108*, 2 (Summer).

Kemp, G. (1991). *The control of the Middle East arms race.* Washington, DC: Brookings Institution Press.

Krause, K. (1997). Cross-cultural dimensions of multilateral non-proliferation and arms control dialogues: An overview. In A. Latham (Ed.), *Non-proliferation agreements, arrangements and responses: Proceedings of the 1996 Canadian non-proliferation workshop*. Toronto: The York Center for International and Security Studies.

Larsen, J. A. (2002). An introduction to arms control. In J. A. Larsen (Ed.), *Arms control: Cooperative security in a changing environment*. Boulder: Lynne Rienner Publishers.

Macintosh, J. (1987). Confidence-building measures—a conceptual exploration. In R. B. Byers, F. S. Larrabee, & A. Lynch (Eds.), *Confidence-building measures and international security*. New York: Institute for East–West, Security Studies.

Macintosh, J. (1996). *Confidence building in the arms control process: A transformation view*. A study prepared for the Non-Proliferation, Arms Control and Disarmament Division Department of Foreign Affairs and International Trade (October).

McFate, P. B. et al. (Eds.), (1994). *The converging roles of arms control verification, confidence-building measures, and peace operations: opportunities for harmonization and synergies*. A Study prepared for the Non-Proliferation, Arms Control and Disarmament Division of the Department of Foreign Affairs and International Trade, Ottawa, Canada (October).

Nachmias, C. F., & Nachmias, D. (1992). *Research methods in the social sciences*. New York: St.Martin's Press.

Reich, B. (1994). The United States and the Arab–Israeli peace process. In G. Ben-Dor & D. B. Dewitt (Eds.), *Confidence building measures in the Middle East*. San Francisco: Westview Press.

Rittberger, V., Efinger, M., & Mendler, M. (1990, February). Toward an east–west security regime: The case of confidence-and security-building measures. *Journal of Peace Research, 27*, 1.

Schelling, T. C., & Halperin, M. H. (1985). *Strategy and arms control*. Washington, DC: Pergamon-Brassy's.

Schofield, J. (2000, December). Arms control failure and the balance of power. *Canadian Journal of Political Science, 33*, 4.

Selltiz, C., Wrightsman, L. S., & Cook, S. W. (1976). *Research methods in social relations*. New York: Holt and Rinehart.

Simon, J. (1978). *Basic research methods in social science and the art of empirical investigation*. New York: Random House.

Steinberg, G. M. (2004). The centrality of confidence-building measures: Lessons from the Middle East. In A. Schnabel & D. Carment (Eds.), *Conflict prevention: From rhetoric to reality*. Ottawa: Lexington Books.

Tanner, F. (2000). The Euro-Mediterranean security partnership: Prospects for arms limitation and confidence-building. In A. Vasconcelos & G. Joffe (Eds.), *The Barcelona process: Building a Euro-Mediterranean regional community*. London: Frank Cass.

Tuchman, B. (1984). *The alternative to arms control.* Los Angeles: Center for International and Strategic Studies, University of California at Los Angeles.

Chapter 3
The Rise of the Arms Control Agenda in the Middle East

With the end of the Cold War, the question of arms control moved to the fore of Middle Eastern politics as an important element in the process of conflict resolution in the region. This upsurge of interest in arms control could be explained in the light of a number of global and regional developments. Perhaps the most important development was the end of the Cold War rivalry between the United States and the Soviet Union. During the Cold War, it was common to dismiss proposals for arms control in the region. The global power relations could hardly have been less appropriate for curtailing the accumulation of weaponry in this troubled area. The Middle East was one of the main battlefields of Soviet–American rivalry, which stimulated a tremendous momentum of conflict in the region, leaving little hope for regional arms control. However, this picture drastically changed with the end of the Cold War, which set the stage for a new chapter in East–West relations as US President Bush declared in his statement of 25 December 1991.[1] The American President declared the beginning of a new era of American–Russian cooperation on global and regional politics. Further, the new Russian leadership pursued a new policy of Euro-Atlanticism according to which it viewed Russia as part of the Euro-Atlantic world and sought more integration with the West. This clearly manifested in the Russian–American Declaration signed at Camp David in February 1992 in which the Russians fully endorsed the American view of international relations.[2] The Camp David declaration reflected an emerging consensus between the United States and Russia on a number of global issues, including the globalization of the concept of arms control and its application to the Middle East. In a letter to the UN Secretary-General on 15 April 1990, the then Soviet foreign minister Edward Shevardnadze

---

[1] See the text of former President Bush's Address to the Nation on the Commonwealth of Independent States, 25 December 1991: available at: http://bushlibrary.tamu.edu/research/public_papers.php?id=3791&year=1991&month=12.

[2] See the text of the American–Russian Camp David Declaration, 1 February 1992; available at: http://www.bits.de/NRANEU/US-Russia/A%20Official%20Docs/Bush%20Yelt%201st%20sum%20.htm.

G. M. Selim, *Global and Regional Approaches to Arms Control in the Middle East*, SpringerBriefs in Environment, Security, Development and Peace 4,
DOI: 10.1007/978-3-642-29314-6_3, © The Author(s) 2013

declared his country's support for controlling the sales of conventional weapons as "a means of building a new model of security" (cited in Neuman 1993: 264). The Russian attitude was further articulated in a statement by former Russian President Yeltsin to the UN Secretary-General on 27 January 1992. In this statement, Yeltsin affirmed his country's commitment to its obligations under the *Nuclear Non-Proliferation Treaty* (NPT) and its support for a global convention on prohibiting chemical and biological weapons (Neuman 1993: 264).

Second, the peaceful end of the Cold War, along with the conclusion of East–West agreements on genuine arms control such as the 1990 Treaty on Conventional Armed Forces in Europe (CFE), raised optimism that the East–West experience of arms control could have applications for other conflict-ridden regions, particularly the Middle East. The argument here was that the East–West arms control regime helped restrain tendencies toward conflict escalation in East–West relations and paved the way for the beginning of a new strategic relationship between the two sides. Accordingly, arms control could also achieve the same task in other conflict-ridden regions, particularly those hosting WMD. In this context, the Middle East received special attention, given the region's substantial record in terms of military conflicts, arms purchases, military spending and the scope of WMD proliferation. This led to the promotion, by many states, international organizations, and security analysts, of arms control as an important instrument that could change the content of Arab–Israeli relations and contribute to the process of conflict prevention and conflict resolution in the Middle East.

In November 1990, the *Charter of Paris for a New Europe*, which ended the Cold War, highlighted the significance of operational arms control and called for developing new generations of CBMs as a means of preventing future conflicts and of building on the peace that resulted from the end of the Cold War. The Charter also outlined a series of strategies and declarations in different regions, emphasizing that the global transformations which led to the end of the Cold War made CBMs the main mechanism for conflict prevention and possibly resolution in other conflict-ridden regions, including the Middle East.[3] In the same vein, Kemp (1991: 164) strongly advocated the application of Western-style arms control in the Middle East. According to his view, "a dialogue on specific security issues under the rubric of arms control might provide a vehicle for more substantive talks on other sources of conflict…In the best of circumstances, arms control talks themselves might be the precursor for wide-ranging peace talks". Moodie (1994: 135) subscribed to a similar view, arguing that arms control is an essential tool for conflict management and conflict resolution in the Middle East (for a similar account, see Feldman 1992: 86).

Third, the question of nuclear proliferation assumed greater urgency at the global and regional levels after the decline of the Soviet nuclear weapons industry and the disintegration of the Soviet Union. This resulted in the break-up of the Soviet nuclear arsenal and the rise of four nuclear successor states (Russia, Ukraine, Kazakhstan, and Belarus). As these states became engulfed in a series of economic and political

[3] *Charter of Paris for a New Europe*, November 1990; available at: http://www.state.gov/t/isn/4721.htm.

upheavals during the early 1990s, concerns were raised about their commitment and ability to control their nuclear arsenals and to cooperate with other nuclear powers in non-proliferation efforts. Given the prevalence of desperate economic conditions in the newly-independent nuclear states, the frightful prospect of these states selling their stockpiles of weapons-grade uranium or plutonium abroad in return for foreign currency became a major concern to the international community. Further, there was the fear that nuclear material could be smuggled into the midst of unstable political conditions. For Western powers, the danger with this potential proliferation was that nuclear materials could reach the hands of renegade states and even terrorist organizations (Landgren 1992: 531–532; Lieber 1995: 5–6).

This was the case, for example, with Iraq when the UN Special Commission on Iraq discovered that Iraq had obtained strategic gyroscopes from dismantled Russian *submarine-launched ballistic missiles* (SLBMs) and had acquired advanced missile technology from Ukraine. There were also reports that Iran was a potential recipient of nuclear-related materials from the newly-independent former Soviet republics. This prompted the United States and other foreign powers to seek precautionary measures to prevent nuclear proliferation. These measures were implemented at two levels; first by cooperating closely with former Soviet republics to tighten up their procedures for ensuring the security of their nuclear arsenals, and second by paying closer attention to the question of arms control and non-proliferation in the Middle East, which, from a US perspective, includes eager purchasers such as Iran, Iraq, Libya, and Syria.

Fourth, the collapse of the Soviet Union had a major impact on the discourse of Arab–Israeli relations. The Arab rejectionist states lost the umbrella of Soviet deterrence, which meant that they could no longer depend on Soviet military or diplomatic support in their political struggle with Israel. In the meantime, Russia pressured its Arab allies, mainly Syria and the *Palestinian Liberation Organization* (PLO), to abandon the cause of armed struggle and to seek a negotiated settlement with Israel. This went in parallel with the readmission of Egypt to the Arab League at the Casablanca Arab summit in 1989 without its being obliged to renounce its 1979 peace treaty with Israel, an implicit acceptance by the Arab states of the legitimacy of negotiation with Israel. This resulted in the moderation of Arab behaviour and the rise of a greater degree of Arab pragmatism toward Israel, which was eventually reflected in the inception of the Madrid Arab–Israeli peace process in October 1991 and later the signing of the Oslo Accords in September 1993 (Lieber 1995: 2). The Madrid and Oslo peace processes created new opportunities for arms control in the Middle East as they opened the political possibility to discuss and negotiate arms control issues in a regional context.[4]

[4] In the past, the Arab rejectionist actors had refused to join any peace talks with Israel. This was mainly a result of their estimation that the talks would mean the recognition of Israel without corresponding Israeli concessions, and because the Soviet Union rejected the first Middle East peace process, which had started in the aftermath of the 1973 war. Meanwhile, the Israelis refused to discuss arms control in isolation from regional security issues. They insisted that any discussion of arms control issues must be made through direct negotiations with their Arab neighbours. See Cohen 1992: 42.

Fifth, the first Gulf War of 1991 had a major impact on the question of arms control in the Middle East. The sudden outbreak of the Gulf crisis in August 1990 demonstrated the extent to which political stability in the region was fragile. In addition, findings published by the UN Special Commission on Iraq and the *International Atomic Energy Agency* (IAEA) in their inspections of Iraq's weapons programme following the war showed that Iraq was much further along in its development of WMD capability than anyone had imagined. This brought to the forefront of the international agenda the issue of the huge stockpiles of arms and the proliferation of WMD, particularly chemical and biological weapons, in the Middle East. In fact, the war set the stage for the removal of the Iraqi WMD arsenal as a prelude to the elimination of all WMD from the Middle East. This linkage was explicitly outlined in UN Security Council Resolution 687, issued in April 1991, which viewed the removal of Iraqi WMDs as "represent[ing] steps towards the goal of establishing in the Middle East a zone free from weapons of mass destruction and all missiles for their delivery and the objective of a global ban on chemical weapons".[5]

More importantly, the Gulf War laid the ground for the emergence of the 'New World Order' and American hegemony in the Middle East. With the end of the war, the United States emerged as the guardian of the Middle East, and embarked upon an ambitious strategy for restructuring the political and military power relationships in the region in line with American interests. In this strategy, arms control was viewed as one of the main restructuring mechanisms. Accordingly, when the Madrid peace process began in 1991, the United States linked the question of arms control with the peace talks, arguing that arms control is a key element for any comprehensive peace settlement in the region. In this context, when the multilateral track of the peace process began, it branched off into five working groups covering the entire spectrum of issues in the Middle East, including arms control. An *Arms Control and Regional Security* (ACRS) working group was established to devise strategies for arms control in the Middle East.

Last but not least, economic factors also created new opportunities for arms control in the Middle East. The oil boom of the 1970s had generated vast revenues for the majority of Arab countries, including the non-oil-exporting ones. Between 1973 and 1982, the Arab countries received approximately US$1 trillion as oil revenues (Sadowski 1994: 29). This caused the defence budget of several Middle Eastern countries to jump from US$25.5 billion in 1975, representing 5.4 per cent of global military expenditure, to US$67 billion in 1980, representing 10.6 per cent of global military expenditure (Al-Fares 1993: 78). During this period, the military expenditure of Middle Eastern countries was considered the largest in the Third World. In the first half of the 1980s, regional military spending further jumped to US$93 billion in 1985, turning the Middle East into the largest market for conventional arms sales in the world (Steinberg 1996: 3–4).

However, the decline of oil prices over the second half of the 1980s had a negative impact on the economies of most Middle Eastern countries. During this period,

[5] See the text of UN Security Council Resolution 687; at: http://www.fas.org/news/un/iraq/sres/sres0687.htm.

Arab oil revenues declined by almost 75 per cent from US$216 billion in 1980 to US$54 billion in 1986 (Sadowski 1994: 30). This led most Middle Eastern countries to reduce their defence budgets, which declined to $53.2 billion by the end of the 1980s (Al-Fares 1993: 78). Whereas the oil-exporting countries began to face limitations on resources available for purchasing weaponry for their own use or the use of their regional allies, the non-oil-exporting countries, particularly Egypt, Syria and Jordan, suffered from growing burdens of debt which substantially affected their military expenditures.[6] In this context, many regional actors began to seek cheaper ways of meeting their defence needs and controlling the costly arms race, and hence to explore the option of arms control.

Taken together, these developments created a momentum for reviving interest in arms control in the Middle East. With the end of the Cold War, the region witnessed a number of ambitious proposals that sought to establish cooperative regional security arrangements in the area of arms control, with a view to limiting the Arab–Israeli arms race and paving the ground for the resolution of the Arab–Israeli conflict. Whereas some of these proposals covered structural arms control, others emphasized the operational dimension of arms control. Some of these proposals were also initiated and discussed at the global level, while others were discussed and debated at the regional level. A detailed review of these proposals is in order.

3.1 The US Arms Control Plan in the Middle East

In the area of structural arms control, a number of global arms control plans were proposed in 1991 in the immediate aftermath of the first Gulf War. The first was the US Arms Control Plan in the Middle East which was delivered by President Bush in May 1991. The American plan was designed against the background of the Gulf War and the threats which accompanied the war with respect to the use of Iraqi surface-to-surface missiles and the potential use of chemical weapons against the US-led coalition and Israel. As a starting point, the Plan defined the Middle East as the region including "Iraq, Iran, Libya, Syria, Egypt, Lebanon, Israel, Jordan, Saudi Arabia, and the other states of the Maghreb and the Gulf Cooperation Council".[7] It may be noticed that Turkey was not addressed by the Bush Plan, and as a result had no arms control obligations under the Plan. The Plan included limitations on the transfer, production and use of three main categories of weapons: (i) conventional weapons, (ii) surface-to-surface missiles, and (iii) WMD.

---

[6] In 1988, Jordan announced it was no longer able to meet its debt obligations, which amounted to US$8 billion. Out of this figure, around US$3 billion was military debt. In 1990, Egypt had a foreign debt of around US$50 billion, of which US$10 billion resulted from arms purchases. Syria also faced a serious debt crisis by the late 1980s. See Steinberg 1996: 8; Sadowski 1994: 36–44.

[7] See the text of the *US Middle East Arms Control Initiative*, 29 May 1991; available at: http://www.disam.dsca.mil/pubs/Vol%2013_4/Fact%20Sheet%20Middle%20East%20Arms%20Control%20Initiative.pdf.

In the area of conventional weapons, the Plan focused on "supplier restraints". It called on the major weapons suppliers, including the five permanent members of the UN Security Council and other members of the G-7, to establish guidelines that put restrictions on "destabilizing transfers of conventional arms" to the Middle East, while simultaneously allowing the countries of the region to "acquire the conventional capabilities they legitimately need to deter and defend against military aggression". It delegated the permanent members of the Security Council to decide collectively the acceptable level of these capabilities. In this context, the Bush Plan listed four guidelines which the permanent members of the Security Council would follow in assessing the acceptable level of conventional capabilities. These were (i) the establishment of effective domestic export controls on the use and exportation of arms transfers, (ii) the exchange of information between the permanent members on arms sales to the Middle East, (iii) the convening of regular meetings to consult on arms transfers, and (iv) the submission of an annual report by each permanent member to the other permanent members on its Middle Eastern arms sales.

In the area of missiles, the American Plan called for a freeze on "the acquisition, production, and testing of surface-to-surface missiles by states in the region with a view to the ultimate elimination of such missiles from their arsenals". It also called on the major weapons suppliers to restrict the export licences of their technology and equipment to their use for peaceful purposes.

In the area of WMD, the Plan called for the establishment of 'a verifiable ban' on the acquisition and production of material used in nuclear weapons such as separated plutonium or enriched uranium, the accession of all Middle Eastern countries to the NPT, the placing of all nuclear facilities in the region under the safeguards of the IAEA, and the eventual creation of a nuclear-free zone in the future. It also called on all countries of the region to adhere to the *Chemical Weapons Convention* (CWC) and the *Biological Weapons Convention* (BWC), and to institute CBMs in the fields of chemical and biological weapons.[8]

3.2 The French Arms Control and Disarmament Plan

Parallel to the Bush Plan, France proposed another arms control plan in the Middle East. This was the Arms Control and Disarmament Plan which was delivered by late French President Mitterrand in June 1991. Similar to the American Plan, the French Plan reflected the growing concern of the major powers with the proliferation of WMD and ballistic missiles in the Middle East and other conflict-ridden regions in the aftermath of the first Gulf War. The French Plan was more comprehensive than its American counterpart. It dealt with three main frameworks for arms control, including (i) the weapons type, (ii) the regional framework, and (iii) the global framework.

[8] Ibid.

Under the weapons type framework, the Plan focused on the ban and elimination of chemical and biological weapons by calling on all regional actors to adhere to the CWC and BWC. It also called for the reduction of existing nuclear weapon arsenals to "the lowest level consistent with the maintenance of deterrence" as well as the non-proliferation of nuclear weapons outside the five nuclear powers. Further, the Plan called for the strict control of the dissemination of ballistic missile technologies to prevent their use for aggressive ends. It advocated the strengthening the *Missile Technology Control Regime* (MTCR) to reduce the risks of ballistic missile technology spreading. In the area of conventional weapons, the Plan called for restraints on the level of conventional arms exports "at the lowest possible level while respecting the right of each state to security", with a view to establishing a balance of power in each region.[9]

Under the regional framework, the Plan made some reference to CBMs by calling on the actors of the region to adopt CBMs in order to "reduce the sense of threat and pave the way for disarmament proper". It called for the creation of an organizational framework—similar to the CSCE framework—under which CBMs would be introduced and applied in the region."The example of Europe is encouraging. Europeans have graduated from cold war to peaceful coexistence and now to cooperation within the framework of the Conference on Security and Cooperation in Europe. They have adopted and applied confidence and security building measures… This approach could serve as an inspiration for other initiatives in the region," the Plan stated.

Under the global framework, the Plan held the UN Security Council responsible for backing and supervising arms control and non-proliferation efforts. "The Council should encourage the signature of agreements on single weapon categories, together with regional and multilateral disarmament and non-proliferation agreements. On the basis of the agreements reached, it could derive general rules and thus maintain a worldwide watch," the Plan stated.

3.3 International Communiqué on Arms Transfers and Non-Proliferation in the Middle East

In the same year, the five permanent members of the Security Council came up with an international communiqué on arms transfers and non-proliferation in the Middle East following their meeting in Paris on 8–9 July. The Communiqué called for the establishment of a WMD-free zone in the Middle East through a comprehensive programme of arms control in the region. This programme would include (i) a freeze on ground-to-ground missiles as a prelude to their complete elimination in the region, (ii) the submission of all nuclear activities in the region to the regular inspection of the IAEA, (iii) a ban on the importation or production of

[9] See the text of the French *Arms Control and Disarmament Plan*, 3 June 1991; available at: http://archives.sipri.org/contents/expcon/mitterand.html.

materials which could be used in producing nuclear weapons, and (iv) the adherence of all states in the region to the CWC.

The five permanent members also called for controlling the process of conventional arms transfers to the region beyond the levels needed for defence purposes. In this context, they agreed to develop "rules of restraint" on conventional arms sales to the region. They also agreed to "develop modalities of consultation and of information exchanges concerning arms transfers to the region". Further, they called upon the states of the region to negotiate regional arms agreements of their own to reinforce policies agreed upon by the international powers. The permanent members also resolved to conduct new talks in September and October of the same year to develop concrete proposals for arms control disarmament in the Middle East.[10]

3.4 The International Declaration on Conventional Arms Transfers to the Middle East

In their October 1991 meeting in London, the five permanent members came up with the International Declaration on Conventional Arms Transfers to the Middle East. The Declaration laid down certain guidelines for the exportation of arms in the Middle East. It stipulated that the permanent members would notify each other of their arms transfers to the countries of the region and that such transfers would occur within the framework of two main guidelines. First, the five powers would consider whether any proposed arms transfer would (i) increase the recipient's ability to exercise legitimate self-defence, (ii) function as a proportionate and appropriate response to the security/military threats facing the recipient country, and (iii) increase the recipient's ability to take part in regional or other collective arrangements or measures consistent with the UN Charter. Second, the five powers would avoid arms transfers in cases where they were likely to (i) intensify an existing arms conflict, (ii) increase regional tension and instability, (iii) be used beyond the legitimate defence needs of the recipient, (iv) be used for intervention in the internal affairs of other sovereign states, and (v) support international terrorism.[11]

3.5 The EU Code of Conduct

In March 1991 the European Council established the Working Group on *Conventional Arms Exports* (COARM) to compare national policies in the area of arms exports and discuss the potential for harmonization. The objective was to

[10] The text of the communiqué is in Anthony et al. 1992: 302–303.

[11] The text of the declaration is in Anthony et al. 1992: 304–305.

develop guidelines for arms transfers under which the EU member states would agree to exercise restraint in exports of conventional arms to other countries. The EU decision was prompted by the Iraqi invasion of Kuwait in 1990 and the scandals involving European countries and companies in the arming of Iraq. The armed conflicts that broke out in south-east Europe and in Africa also led European governments to consider the relationship between arms exports and issues of economic development, human rights, and conflict prevention. In response to these concerns, the Luxembourg and Lisbon European Councils in 1991 and 1992 adopted eight criteria against which EU member states agreed to assess their arms exports. These criteria drew in large part on the previously outlined guidelines drawn up in 1991 by the five permanent members of the UN Security Council. These included:

1. Respect for the international obligations and commitments of member states, in particular the sanctions adopted by the United Nations Security Council or the European Union, and agreements on non-proliferation and other subjects, as well as other international obligations.
2. Respect for human rights in the country of final destination as well as respect by that country for international humanitarian law.
3. Internal situation in the country of final destination, as a function of the existence of tensions or armed conflicts.
4. Preservation of regional peace, security and stability.
5. National security of member states and of territories whose external relations are the responsibility of a member state as well as that of friendly and allied countries.
6. Behaviour of the buyer country with regard to the international community, in particular its attitude to terrorism, the nature of its alliances and its respect for international law.
7. Existence of a risk that the military technology or equipment will be diverted within the buyer country or re-exported under undesirable conditions.
8. Compatibility of the exports of the military technology or equipment with the technical and economic capacity of the recipient country, taking into account the desirability that states should meet their legitimate security and defence needs with the least diversion of human and economic resources for armaments.[12]

In the Cardiff European Council of June 1998, the EU member states adopted an EU Code of Conduct on Arms Exports, with the stated objective to "prevent the export of equipment which might be used for internal repression or international aggression or contribute to regional instability".[13] Under the EU Code of Conduct,

---

[12] "Presidency Conclusions". European Council, Luxembourg, 28–29 June 1991; available at: http://aei.pitt.edu/1935/1/1935.pdf; and European Council, Lisbon, 26–27 June 1992; available at: http://aei.pitt.edu/1420/1/Lisbon_june_1992.pdf.

[13] Council of the European Union, 'European Union Code of Conduct on Arms Exports', Doc. 8675/2/98 REV 2, 5 June 1998.

the EU member states committed themselves to set "high common standards which should be regarded as the minimum for the management of, and restraint in, conventional arms transfers" and "to reinforce cooperation and to promote convergence in the field of conventional arms exports" within the framework of the EU Common Foreign and Security Policy.[14] The EU Code further elaborated the eight criteria adopted in the Luxembourg and Lisbon summits of 1991 and 1992. The Code required member states to deny an export licence if the transfer was deemed to conflict with any of criteria 1–4 and to take into account the factors listed in criteria 5–8 when considering a licence application. It also outlined reporting procedures and consultation mechanisms intended to ensure more consistent interpretation of the criteria by member states. Further, the EU member states agreed to exchange confidential information on their denials of arms export licences along with aggregated data on their export licence approvals and their actual exports. They also agreed to consult other member states when considering the granting of an export licence identical to a licence that another member state had denied within the previous 3 years. The data on licences and exports would be compiled in the publicly available annual report according to Operative Provision 8 of the EU Code.

In December 2008, the EU Code of Conduct was replaced with the EU Common Position, which formally identified the range of activities that should be covered by member states' arms export licensing systems. In addition to 'physical exports', member states' arms export licensing systems should also cover licensed production; brokering; transit and trans-shipment; and intangible transfers of software and technology. Other changes included modifications to the eight criteria for further restrictions on conventional arms transfers.[15]

3.6 The ACRS Working Group

Parallel with global proposals on arms control in the Middle East, the question of arms control was addressed at the regional level within the framework of the ACRS working group. The Madrid Peace Conference of October 1991 branched off into two main tracks: bilateral and multilateral. The bilateral track dealt with bilateral Arab–Israeli (Palestinian–Israeli, Jordanian–Israeli, and Syrian–Israeli) negotiations to reach a political settlement of the question of the Arab territories occupied in 1967. The multilateral track entailed discussions among the regional and global actors to introduce measures that would support the political settlement. In these discussions, some regional actors did not participate, either because they refused, such as Syria and Lebanon, or because they were not invited, such as

[14] Ibid.

[15] Council Common Position 2008/944/CFSP of 8 December 2008 defining common rules governing control of exports of military technology and equipment, *Official Journal of the European Union*, L335, 8 December 2008.

Iraq, Iran, and Libya. Meanwhile, most of the big powers were active participants in these discussions. The multilateral talks took place in five working groups covering key issues that concerned the entire Middle East, including water, environment, refugees, economic development, and arms control.

The ACRS was the group in which the issue of arms control was discussed. The ACRS sessions were chaired by the United States and Russia. However, it was the United States that took the leading role in the process, given that Russia was going through major domestic upheavals which constrained its international role. This was in addition to the fact that Yelstin, the then Russian President, was pursuing a Euro-Atlantic policy that was directed towards integration with the West and endorsement of American global and regional policies. In fact, Russia left the entire process to be managed by US negotiators.

Two main 'baskets' were developed within the context of the ACRS working group. The first was the conceptual basket. In this basket, the long-term objectives of arms control were discussed. In May 1994, a meeting of the ACRS was held in Cairo to draw up a draft declaration on regional security and arms control in the Middle East. This draft was discussed in the Cairo meeting, but was not endorsed. The second was the operational basket, which dealt with the question of CBMs, such as the exchange of information on militaries, pre-notification of military exercises, the creation of a network, and joint research and rescue exercises at sea. In December 1994, the participants at the ACRS reached an important CBM when they accepted that plans to move more than 4,000 troops or 110 tanks would be notified in advance. They also made significant progress on the exchange of military officers, information, unclassified military publications, structures of defence forces, and curricula for the education of senior military officers. At a meeting held in Amman in November 1995, the parties agreed on setting up a regional crisis prevention centre. Among the initial steps towards the establishment of this centre was the undertaking by the parties to provide information about their military arsenals for a databank on arms in the Middle East. In addition, the ACRS participants agreed on two framework agreements in the field of maritime CBMs. These were (i) a regional 'Prevention of Incidents at Sea' agreement, and (ii) a regional 'Search and Rescue Operations' agreement. Indeed, the discussion of maritime CBMs was considered the most successful panel of the ACRS. Progress was also made on search and rescue operations and on the prevention of incidents at sea (Jones 1997: 65–70).

3.7 The NATO–Mediterranean Dialogue

The initial success of the ACRS discussions in the operational basket and the development of CBMs proposals helped to direct global attention to the question of confidence-building as a prelude to structural arms control in the Middle East. This motivated Western powers to give more emphasis to CBMs as an alternative way of approaching the question of arms control in the Middle East.

It was in this context that NATO presented its first arms control initiative in the region in 1995 under the name of the NATO–Mediterranean Dialogue. In this year, NATO suggested the establishment of a dialogue with the southern Mediterranean countries on CBMs. The idea of the NATO–Mediterranean Dialogue was first articulated at the Athens ministerial meeting of the *North Atlantic Council* (NAC) of NATO in June 1993 as well as the Brussels NATO Summit in January 1994 when NATO leaders expressed concern about the impact of developments in the Mediterranean region on European security. With the signing of the Israeli–Palestinian *Declaration of Principles* (DOP) in 1993, NATO leaders expressed their readiness to contribute to the reinforcement of peace and security in the Mediterranean through the application of some form of CBMs in the region. On 8 February 1995, the NAC decided to initiate a direct dialogue with Egypt, Morocco, Tunisia, Israel, and Mauritania. The first meeting of the Dialogue was held in Brussels between the General Secretariat of NATO and the embassies of these countries. Libya, Syria, and Lebanon were absent from the Dialogue, which was held in a bilateral and undeclared framework. In December 1995, Jordan was incorporated in the Dialogue, and Algeria participated in February 2000 (Attina 2001: 39).

In a seminar held in Rome in 1996, Balazino, NATO's deputy secretary general, asserted that the objective of the dialogue was to build a series of political relations which could provide each of the participating states with a sense of security. Balazino also made it clear that NATO would not take part in the processes of conflict resolution and structural arms control in the region, but would rather focus on issues related to confidence-building, including (i) the exchange of military information, (ii) the conduct of joint limited military exercises, and (iii) the holding of joint seminars on security-related questions. In November 2000, the NATO summit suggested a series of CBMs with the aim of establishing a cooperative security regime between NATO, Israel, and the Mediterranean Arab countries. These included (i) military education, (ii) training to address basic interoperability requirements with a view to making Mediterranean countries better prepared to participate in military exercises and related training activities, (iii) military medicine, defence reform, and civilian management of defence forces, (iv) fighting terrorism, (v) border security, especially in connection with terrorism, (vi) smuggling of small arms and other illegal activities, (vii) civil emergency planning including disaster management, and (viii) science and environment including activities in the field of desertification, management of water and other natural resources, and environment pollution (Larrabee et al. 1998: 45–49).

3.8 The Euro-Mediterranean Partnership

In addition to NATO, the EU also came up with its own CBMs proposals in the Middle East. In November 1995, the EU formally presented the Euro-Mediterranean Partnership (EMP) initiative as a framework for cooperation with 12 Mediterranean countries, including 8 Arab countries (Morocco, Algeria, Tunisia, Egypt, Jordan, the

Palestinian Authority, Syria, and Lebanon), in addition to Israel, Cyprus, Malta, and Turkey. The proposed EMP had three major objectives. The first was to establish a Euro-Mediterranean Zone of Political Stability and Security through political dialogue between the EU and the Mediterranean countries. This dialogue was based on the principles of democracy, good governance, the rule of law, and the efforts to persuade these countries to renounce non-conventional military options. A code of conduct for the resolution of disputes was also suggested. The second was to establish an area of shared prosperity through the establishment of a Euro-Mediterranean Economic Area in which all manufactured products would be traded freely. The third was to promote cultural understanding and cooperation between the two shores of the Mediterranean.[16]

In October 1995, the EU issued a statement that operationalized its EMP initiative. The policy statement outlined three main aspects of the EMP: a political-security aspect, an economic financial aspect, and a social human aspect. Upon the EU initiative, the first ministerial meeting of the 27 Euro-Mediterranean partners was held in Barcelona to establish a framework for cooperation in the Mediterranean region. This resulted in the labelling of the entire process as the Barcelona Process, encompassing 15 EU member states, in addition to Egypt, Algeria, Morocco, Tunisia, Jordan, Syria, Lebanon, Cyprus, Israel, Malta, Turkey, and the Palestinian Authority (Derisbourg 1997: 10–12).

In the Barcelona meeting, the EU advocated the application of CBMs within the framework of the EMP. This was articulated in the Barcelona Declaration, which stated that the actors would "consider any confidence and security-building measures that would be taken between the parties with a view to the creation of an area of peace and stability in the Mediterranean, including the long term possibility of establishing a Euro-Mediterranean pact to that end". The formulation of CBMs within the EMP was the subject of follow-up discussions about the main parameters of the EMP. The proposed CBMs included the reporting of military expenditures, prior notification of major military manoeuvres, exchange of information on defence activities, exchange of expertise at the official level on conflict management, and cooperation in peacekeeping missions. In addition, other measures were suggested such as cooperation in natural crisis management, and air and sea search and rescue operations (Asseburg 2003: 174–180).

In 1999, the EU extended its CBMs proposal in the Mediterranean region by drafting the Charter for Peace and Stability in the Mediterranean. The Conclusions of the Euro-Med Ministerial Conference in Stuttgart defined the proposed Charter

[16] The Political-Security Basket aimed at establishing a common area of peace and stability in the Mediterranean based on the principles of human rights and democracy. This included the adoption of measures concerning respect for democracy and the rule of law, human rights, the rights of self-determination, non-interference in the internal affairs of other states, cooperative measures for countering terrorism, and the settlement of disputes by peaceful means. The EMP also adopted soft security practices, including regular political and security dialogues, security expert meetings, seminar diplomacy, and partnership-building measures with the aim of creating trust and collective security understandings between EU members and partner states. The Economic and Financial Basket aimed at creating a Euro-Mediterranean free trade area by 2010. The EU would provide financial aid to the southern Mediterranean partners to develop the private sector and promote structural economic reforms. The Cultural and Human Basket sought to establish a partnership in cultural and social affairs.

as the "instrument for the implementation of the principles of the Barcelona Declaration where issues of peace and stability are concerned". For this purpose, the Charter would "provide for an enhanced political dialogue as well as the evolutionary and progressive development of partnership-building measures, good-neighbourly relations, regional cooperation and preventive diplomacy". The primary function of the enhanced political dialogue would be "to prevent tensions and crises and to maintain peace and stability by means of cooperative security". The proposed Charter referred explicitly to a set of CBMs in the Mediterranean, and its objective was to establish a set of security guidelines which would be observed by the Mediterranean countries under the supervision of the EU. The main focus of the Charter was in the field of 'conflict prevention'.[17] This meant that the proposed Charter would employ CBMs that were necessary for preventing the outbreak of future conflicts between the Mediterranean actors.

References

Al-Fares, A. R. (1993). *Arms and bread: Arms expenditure in the Arab World.* Beirut: Center for Unity Studies (in Arabic).

Anthony, I., Allebeck, A. C., Miggiano, P., Sköns, E., & Wulf, H. (1992). The trade in major conventional weapons. In *SIPRI yearbook 1992: World armament and disarmament.* Oxford: Oxford University Press.

Asseburg, M. (2003). The EU and the Middle East conflict: Tackling the main obstacle to the euro-mediterranean partnership. *Mediterranean Politics, 8*, 2.

Attina, F., & Attina, F. (2001). Partnership and security: Some theoretical and empirical reasons for positive developments in the euro-mediterranean area. In F. Attina & S. Stavridis (Eds.), *The Barcelona process and euro-mediterranean issues from Stuttgart to Marseilles*. Milano: Dott. A. Giuffre Editors.

Cohen, A. (1992). Patterns of nuclear opacity in the Middle East: Understanding the past, implications for the future. In T. Rauf (Ed.), *Regional approaches to curbing nuclear proliferation in the Middle East and South Asia*. Ottawa: Canadian Center for Global Security.

Derisbourg, J. P. (1997). The Euro-Mediterranean partnership since Barcelona. In Gillespie, R. (Ed.), *The Euro-Mediterranean partnership: Political and economic perspectives.* London: Frank Cass.

Feldman, S. (1992). Security and arms control in the Middle East: An Israeli perspective. In S. A. Stahl & G. Kemp (Eds.), *Arms control and weapons proliferation in the Middle East and South Asia*. New York: St. Martin's Press.

Jones, P. (1997). Arms control in the Middle East: Some reflections on ACRS. *Security Dialogue, 28*, 1.

Kemp, G. (1991). *The control of the Middle East arms race.* Washington, DC: Brookings Institution Press.

Landgren, S. (1992). Post soviet threats to security. In *SIPRI Yearbook 1992: World Armament and Disarmament.* Oxford: Oxford University Press.

Larrabee, S. F., Green, J., Lesser, I. O., & Zanini, M. (1998). *NATO's Mediterranean initiative: Policy issues and dynamics*. Washington, DC: Rand.

---

[17] Third Euro-Mediterranean Conference of Foreign Ministers Stuttgart, 15–16 April 1999. Available online at: http://www.eeas.europa.eu/euromed/conf/stutg_conc_en.pdf.

Lieber, R. J. (1995). American hegemony, regional security and proliferation in the post-cold war international system. In E. Inbar & S. Sandler (Eds.), *Middle Eastern security: Prospects for an arms control regime*. London: Frank Cass.

Moodie, M. (1994, September). Constraining conventional arms transfers. *Annals of the American Academy of Political and Social Science*, 535, The Arms Trade: Problems and Prospects in the Post-Cold War World.

Neuman, S. (1993). Controlling the arms trade: Prospects for the future. In E. Karsh, M. S. Navias, & P. Sabin (Eds.), *Non-conventional weapons proliferation in the Middle East: Tackling the spread of nuclear, chemical, and biological capabilities*. Oxford: Clarendon Press.

Sadowski, Y. M. (1994). *Scuds or butter? The political economy of arms control in the Middle East.* Cairo: Al-Ahram Foundation (in Arabic).

Steinberg, G. (1996). Opportunities for conventional arms limitations in the Middle East and Persian Gulf. In *Paper for the Conference of Conventional Arms Transfers*, sponsored by the Carnegie Endowment for International Peace, 12 January.

Chapter 4
The Post-9/11 Era: Is Arms Control Still Relevant?

Following the 9/11 attacks, a major change occurred in Western, particularly American security policies at the global and regional levels, including policy in the area of arms control. In the United States, the George W. Bush administration declared a war on terrorism that targeted both terrorist organizations and so-called 'rogue states'. The administration argued that it could not afford to wait until such organizations or states acquired WMDs and used them first against US interests. Accordingly, the United States had to be prepared to strike first against aspiring possessors of WMD, especially rogue states seeking nuclear weapons. This brought the war on terrorism into a war of counter-proliferation in which the United States and a number of its European allies would target non-state terrorist organizations and rogue states suspected of developing WMDs (Record 2004).

4.1 The Rise of the Counter-Proliferation Doctrine

With the advent of the Bush administration in 2001, the US conception of arms control was seriously re-examined. The administration downgraded the utility of arms control as a tool for combating the proliferation of WMD and protecting US national security, arguing that "multilateral arms control treaties helped to limit proliferation and provide an international basis for rallying political coalitions, but, by themselves, could not stop a determined proliferator, such as Iraq or North Korea".[1] It also downplayed the significance of traditional 'deterrence', which constituted an integral part of the arms control thinking during and after the Cold War. In June 2002, President Bush made it clear that "deterrence means nothing against shadowy terrorist networks with no nations or citizens to defend. Containment is not possible when unbalanced dictators with weapons of mass destruction can deliver those weapons on missiles or secretly provide them to terrorist allies". He added, "We

[1] *IISS Strategic Survey 2001/2002*, p. 20.

G. M. Selim, *Global and Regional Approaches to Arms Control in the Middle East*, SpringerBriefs in Environment, Security, Development and Peace 4,
DOI: 10.1007/978-3-642-29314-6_4, © The Author(s) 2013

cannot defend America and our friends by hoping for the best. We cannot put our faith in the word of tyrants, who solemnly sign non-proliferation treaties, and then systemically break them. If we wait for threats to fully materialize, we will have waited too long".[2] In this context, arms control became increasingly seen by US officials as "ponderous at best and counterproductive at worst" (Newman 2004: 59). The result was the rise of a new security doctrine that centred on the principles of pre-emption and "muscular counter-proliferation", according to which the United States would seek not just to prevent the spread of WMDs, but also to roll back and eliminate, by force if necessary, such weapons from the arsenals of rogue states suspected of developing WMD (Litwak 2007: 62).

Although the concept of counter-proliferation was not uncommon in Western security literature in the post-Cold War era, it did not gain prominence within US official policy and defence circles prior to 9/11. In 1993, the concept was officially named in the US security doctrine with the launching of the Defense Counterproliferation Initiative (DCI) by the then US Secretary of Defence Les Aspin. The DCI centered on developing new weapons to destroy WMD; re-examining the strategies used against the new kind of threat; directing intelligence efforts to detect WMD; and ensuring international cooperation in curtailing the threat of such weapons. It was later integrated in the 1999 National Security Strategy which emphasized the U.S. commitment to non-proliferation efforts, with military counter-proliferation efforts in a support role.[3] During this period, counter-proliferation was viewed primarily as "a conventional counter-force challenge: the threatened or actual use of nonnuclear weapons to deter or prevent a nuclear adversary from using nuclear weapons" (Record 2004: 8). In this respect, counter-proliferation was not extended to include the use of pre-emptive strikes aimed at preventing a regime from acquiring WMD, and it remained confined within the boundaries of international law, which "prohibits military strikes against states not at war except in circumstances of imminent and indisputable enemy attack" (Record 2004: 8). In a transformed post-9/11 global order, the conception of counter-proliferation moved from the shadows to the forefront of US security policy, and even acquired more aggressive dimensions.

It is important to note, however, that the Bush doctrine of counter-proliferation and pre-emption was not merely an outcome of the 9/11 attacks as some observers might claim. While it is true that the new doctrine was officially outlined in the aftermath of 9/11, its core elements and principles had been articulated 4 years earlier with the foundation of the organization of the *Project for the New American Century* (PNAC), an American think tank established in 1997 by a group of neoconservatives who wanted to express their views on American global leadership and eventually took up key positions in the Bush administration in 2001. In September 2000, a few months before the Bush administration assumed office, the

---

[2] Remarks at West Point commencement ceremonies, 1 June 2002.

[3] Secretary of Defense Les Aspin, speech to the National Academy of Sciences. Washington, D.C., 7 December 1993.

PNAC issued a document entitled *Rebuilding America's Defenses: Strategy, Forces and Resources For the New Century*. The document supported a "blueprint for maintaining global U.S. preeminence, precluding the rise of a great power rival, and shaping the international security order in line with American principles and interests". This "American grand strategy" must be advanced for "as far into the future as possible", the document stated. The document acknowledged that "preserving the desirable strategic situation in which the United States now finds itself requires a globally preeminent military capability both today and in the future". It described American armed forces abroad as "the cavalry on the new American frontier". The document also adopted a unilateralist posture by the United States in the world and cast aside international legality and obligations by stressing that "in no circumstances should America's politics be crippled by the misguided insistence of the Security Council on unanimity".[4] As far as arms control was concerned, Schmitt (2001) from the PNAC explained, "Conservatives do not like arms control agreements for the simple reason that they rarely, if ever, increase U.S. security.... The real issue here, and the underlying question, is whether the decades-long effort to control the proliferation of weapons of mass destruction and the means to deliver them through arms control treaties has in fact worked." He contended that it was no longer "plausible to argue that our overall security was best served by a web of parchment accords, and not our own military capabilities".

Although many neoconservatives assumed high government positions in the George W. Bush administration, it was not until the 9/11 attacks that they were able to profoundly change the course of US arms control and non-proliferation policy. In 2002, the main elements of the Bush doctrine were officially articulated with the publication of the *Nuclear Posture Review* (NPR), the *National Security Strategy* (NSS), and the *National Strategy to Combat Weapons of Mass Destruction* (NSCWMD) in February, September, and December of that year respectively.

The NPR is a classified document that is mandated by law, and produced periodically by the US Department of Defense. In the version of the NPR which was leaked by the *Los Angeles Times* in March 2002, the document envisioned the creation of a more diverse set of options for deterring 'rogue states' with the potential to threaten the United States with WMD. It undermined deterrence and the normative taboo on the use of nuclear weapons by identifying a variety of ways in which nuclear weapons could help the United States achieve its goals of assuring allies, dissuading adversaries from acquiring capabilities that threaten the United States, deterring attacks against the United States and its allies, and defeating adversaries. Indeed, the NPR called for the United States to be prepared to use nuclear weapons in a wide spectrum of scenarios, ranging from a conflict with China (possibly over Taiwan) to an Arab–Israeli conflict and/or a conflict with an 'emerging threat' such as North Korea, Iran, or Syria. It also called for developing a more flexible

[4] *Rebuilding America's Defenses: Strategy, Forces and Resources For a New Century*. A document prepared by the Project for the New American Century, September 2000; available at: <http://www.newamericancentury.org/RebuildingAmericasDefenses.pdf>.

targeting system that would enable the United States to quickly prepare nuclear attack options for unforeseen contingencies. According to the NPR, "current conventional weapons are not effective for the long-term physical destruction of deep, underground facilities", and current nuclear weapons, which have limited ground penetration capability, do not "provide a high probability of defeat of these important targets". With the development of a new generation of smaller but more effective nuclear weapons such as 'mini-nukes' and 'bunker busters', "many buried targets could be attacked using a weapon with a much lower yield…[which] would achieve the same damage while producing less fallout (by a factor of 10–20) than would the much larger yield surface burst", the document stated.[5]

The NSS marked the elevation of pre-emption to official US security doctrine. The document is a succinct presentation of a strategy of military dominance that rejected the policies of deterrence, containment, and collective security. Instead, the new strategy stressed offensive military intervention, pre-emptive first strikes, and proactive counter-proliferation measures against rogue states. According to this grand scheme, US security strategy no longer depended on defence and reaction, but on offensive measures, and the only path to security was the path of military action. According to the NSS:

- The gravest danger our Nation faces lies at the crossroads of radicalism and technology. Our enemies have openly declared that they are seeking WMD, and evidence indicates that they are doing so with determination.
- The US will launch pre-emptive strikes against countries suspected of developing weapons of mass destruction (WMD) and which are perceived as threat to its security; it will also attack, disrupt and destroy the terrorist organizations that have a global reach before they penetrate its borders or those of its allies, in what it called 'pro-active counter-proliferation'.
- We will ensure that all needed capabilities to combat WMD are fully integrated into the emerging defense transformation plan and into our homeland security posture. Counter proliferation will also be fully integrated into the basic doctrine, training, and equipping of all forces to ensure they can sustain operations to decisively defeat WMD-armed adversaries.
- Because deterrence may not succeed, and because of the potentially devastating consequences of WMD use against our forces and civilian population, US military forces and appropriate civilian agencies must have the capability to defend against WMD-armed adversaries, including in appropriate cases through pre-emptive measures. This requires capabilities to detect and destroy an adversary's WMD assets before these weapons are used.
- The United States will always proceed deliberately, weighing the consequences of its actions. To support pre-emptive options, the United States will coordinate closely with its closest allies to form a common assessment of the most danger-

[5] Nuclear Posture Review [Excerpts]. Submitted to Congress on 31 December 2001; available at <http://www.stanford.edu/class/polisci211z/2.6/NPR2001leaked.pdf>.

ous threats, and will continue to transform its military forces to ensure US ability to conduct rapid and precise operations to achieve decisive results.[6]

The NSCWMD reiterated the main pillars of the Bush doctrine. The document began by affirming that "As with the war on terrorism, our strategy for homeland security, and our new concept of deterrence, the U.S. approach to combat WMD represents a fundamental change from the past." It declared that rogue states do not regard WMD as weapons of last resort, but rather as "militarily useful weapons of choice intended to overcome our nation's advantages in conventional forces and to deter us from responding to aggression against our friends and allies in regions of vital interest".[7]

In addition, the NSCWMD outlined three principal pillars of the US strategy for combating WMDs. These included, (i) counter-proliferation, according to which the United States "must enhance the capabilities of its military, intelligence, technical, and law enforcement communities to prevent the movement of WMD materials, technology, and expertise to hostile states and terrorist organizations". Under this pillar, the United States would also pursue a strategy of active defence, according to which "U.S. military forces and appropriate civilian agencies must have the capability to defend against WMD-armed adversaries, including in appropriate cases through preemptive measures". This requires capabilities to detect and destroy an adversary's WMD before these weapons are used; (ii) non-proliferation, according to which the United States must "dissuade supplier states from cooperating with proliferant states and induce proliferant states to end their WMD and missile programs, hold countries responsible for complying with their commitments, continue to build coalitions to support these efforts, as well as to seek their increased support for non-proliferation and threat reduction cooperation programs". Non-proliferation efforts might also include the strengthening of US export controls and the development of a comprehensive sanctions policy. However, the document emphasized that should non-proliferation efforts fail, the United States must have full operational and military capabilities necessary to disarm rogue states and terrorist organizations; and (iii) WMD consequence management, according to which the United States "must be fully prepared to respond to the consequences of WMD use on its soil, whether by hostile states or by terrorists…and to respond to the effects of WMD use against its forces deployed abroad, and to assist friends and allies".[8]

All of this was not very new, but the accent and the ways the Bush administration implemented this strategy were novel. The documents, in fact, formalized the ad hoc policy decisions of previous US administrations. Pre-emptive actions against real or alleged foreign enemies and threats had been used in US history, but this was the first time that such operations became part of the US official doctrine. In the name of US national interests, the Bush administration cancelled

[6] *The National Security Strategy of the United States of America*, September 2002; available at: <http://merln.ndu.edu/whitepapers/USnss2002.pdf>.

[7] *National Strategy to Combat Weapons of Mass Destruction*, December 2002; available at: <http://www.fas.org/irp/offdocs/nspd/nspd-wmd.pdf>.

[8] Ibid.

previous prohibitions on the physical elimination of selected foreign statesmen accused of plotting against the US, or of some other wrongdoing. This was part of the new 'regime change' policy targeting 'rogue states', which were handpicked by Washington's policymakers because they were allegedly developing WMD, or supporting terrorism, or both. The new policy of pre-emption was not elevated, however, to the rank of a universal tool available to everybody. It was supposed to remain a US global prerogative, but allowing the possibility of regional pre-emptive strikes by hand-picked strategic allies, such as Israel (Dokos 2008: 111–120).

It was in this context that the United States proposed the *Proliferation Security Initiative* (PSI) in May 2003 as a proactive tool to counter the proliferation of WMD by both state and non-state actors. In addition to the United States, the PSI was joined by ten other countries, namely France, Britain, Germany, Italy, Japan, Netherlands, Australia, Poland, Portugal and Spain. In September 2003, the PSI participants approved a series of interdiction principles that aimed to stop shipments of biological, chemical, and nuclear weapons, as well as missiles and goods that could be used to deliver or produce such weapons, to terrorists and countries suspected of trying to acquire WMD capabilities. This was to be achieved by carrying out cargo interdictions, by force if necessary, for shipments of goods believed to be part of illicit WMD programmes.[9]

It is revealing that the actual application of the PSI did not cover all states outside the nuclear club, and which were believed to have WMD capabilities. Rather, the PSI was limited to stopping shipments of WMD and dual-use material that had both civilian and military purposes to those countries and non-state actors that were viewed as 'threats' by the United States and other major PSI participants. This selective application of the PSI, which was mainly driven by political predilections, was made clear when John Bolton, the then US Under Secretary of State for Arms Control and International Security, and a chief architect of PSI, indicated in November 2003 that PSI participants would not be targeting the trade of countries perceived as US allies or friends, such as India, Israel, and Pakistan—all three of whom possess WMD arsenals, including nuclear weapons.[10] The PSI was thus greeted with a high degree of scepticism by the majority of states outside the PSI. Although the initiative did not grant its participants any new legal authority to conduct interdictions in international waters or airspace, the fact that there was no formal treaty defining the PSI activities or the list of criteria by which interdictions were to be made raised legitimate concerns about the extent to which the PSI was consistent with international law.[11]

In the same vein, France made significant transformations in its nuclear deterrence doctrine. The revised doctrine was announced on 19 January 2006 in an address by the then president Jacques Chirac at the nuclear headquarters of the Strategic Air and Maritime Forces in Brittany, France. The doctrine, which was in

[9] US Department of State, "Proliferation Security Initiative," May 2003; available at: <http://www.state.gov/t/isn/c10390.htm>.

[10] "The Proliferation Security Initiative: An Interview with John Bolton." *Arms Control Today*, December 2003; available at: <http://www.armscontrol.org/act/2003_12/PSI>.

[11] For a review of the legal concerns pertaining to the PSI, see Khurana 2004: 239–241.

line with the 2002 American NPR, emphasized that France's nuclear arsenal would continue to defend the country's vital interests, but it broadened the definition of those interests beyond traditional concerns such as the protection of territory and population and the free exercise of sovereignty. The doctrine identified France's new vital interests as strategic supplies, the defence of allied countries, and even threats or blackmail against these interests, all of which could require a nuclear response from Paris. In addition, the new doctrine expanded the list of countries to be deterred by the French nuclear arsenal to include states that supported terrorists. According to Chirac, "The leaders of states who would use terrorist means against us, as well as those who would consider using, in one way or another, weapons of mass destruction, must understand that they would lay themselves open to a firm and adapted response on our part…This response could be a conventional one. It could also be of a different kind." Chirac added that in responding to threats from regional powers, French strategic nuclear weapons were reconfigured to be more flexible and reactive, enabling Paris to respond directly to such powers.[12]

At a technical level, the new French doctrine emphasized the development of a new generation of smaller but more effective nuclear weapons that could be used against specific targets. Similar to US mini-nukes, France developed a number of SLBMs that could carry a smaller number of warheads for tactical use against selected targets. According to Yost (2006: 704), "This would enable France to undertake what is called a 'split launch', which refers to the launching of one or a few missiles instead of the entire boatload of 16 missiles, as had been France's policy during the Cold War and beyond". He added that "An SLBM equipped with only one or two nuclear warheads could cause much less destruction than one armed with six warheads, particularly if these warheads were delivered with great accuracy. … An SLBM with fewer warheads would have greater range options".

Similarly, Britain upgraded its nuclear doctrine in the aftermath of 9/11. In December 2006, the British government issued a White Paper entitled *The Future of the United Kingdom's Nuclear Deterrent* in which it assigned a greater role to its nuclear weapons in dealing with a wide variety of threats to British national security and interests. The 2006 White Paper outlined four broad areas in which the logic of nuclear deterrence was justified.

The first maintained that nuclear weapons were required to provide a deterrent against the re-emergence of "a major direct nuclear threat to the UK or our NATO Allies" and to prevent major wars that might threaten the survival of the nation. The second area focused on deterring the use of nuclear weapons or other WMD by so-called 'rogue' states in the context of regional intervention. The document stressed that as Britain would continue to intervene in regional crises with conventional military forces, it would need to keep its nuclear weapons in order to deter a

[12] "Our concept for the use of nuclear weapons remains unchanged, President Jacques Chirac speech on French nuclear doctrine." *Acronym Institute for Disarmament Policy*, 19 January 2006; available at: <http://www.acronym.org.uk/proliferation-challenges/nuclear-weapons-possessors/france/our-concept-use-nuclear-weapons-remains-unchanged-president-jacques-c?page=show>.

'rogue' state from using its WMD against Britain or its vital interests. In this context, nuclear weapons provided "an assurance that we cannot be subjected in future to nuclear blackmail" or coercion by a 'rogue' state attempting to "deter us and the international community from taking the action required to maintain regional and global security". The third area focused on deterring "state-sponsored acts of nuclear terrorism". The document stated that "There are risks that international terrorists may try to acquire nuclear weapons. While our nuclear deterrent is not designed to deter non-state actors, it should influence the decision-making of any state that might consider transferring nuclear weapons or nuclear technology to terrorists." The fourth area focused on providing a general deterrent in an uncertain future characterized by (i) the further spread of nuclear weapons; (ii) a likely increase in complex, regional conflicts that could threaten Britain's vital interests; and (iii) the risk of future 'strategic shocks' that could undermine Britain's security.[13]

Although the concept of 'vital interests' was not defined in the document, it had been delineated in a 2003 White Paper entitled *Delivering Security in a Changing World* to include the deterrence of threats to the security of the European continent, global economic interests based on the free flow of trade, overseas and foreign investment and key raw materials, the safety and security of British citizens living and working overseas and its overseas territories, and general international stability.[14] This broad concept of 'vital interests' went beyond threats to the survival of the nation, and extended the domain of nuclear weapons from serving only as a deterrent and weapon of last resort to other areas where they could be employed in situations that fall short of a total war waged for absolute survival. This opened the door to the tactical use of these weapons in response to conventional threats, especially if the attacker perceived no deterrent.

In 2010, the British government issued another document entitled *Securing Britain in an Age of Uncertainty: The Strategic Defence and Security Review*, in which it further articulated its defence and nuclear deterrence posture. The document provided assurances that Britain "will not use or threaten to use nuclear weapons against non-nuclear weapon states parties to the NPT." However, it emphasized that "this assurance would not apply to any state in material breach of its non-proliferation obligations". The document did not delineate specific criteria for determining when a suspected actor is said to be in violation of its non-proliferation commitments. Rather, the statement was ambiguously framed, leaving the door open for the British government to use its nuclear weapons in line with its own assessment of other states' commitment to their nuclear obligations. The document added that "While there is currently no direct threat to the UK or its vital interests from states developing capabilities in other weapons of mass destruction, for

[13] *The Future of the United Kingdom's Nuclear Deterrent*, Defence White Paper, December 2006; available at: <http://www.fas.org/nuke/guide/uk/doctrine/sdr06/WhitePaper.pdf>.

[14] *Delivering Security in a Changing World*, Defence White Paper, December 2003; available at: <http://merln.ndu.edu/whitepapers/UnitedKingdom-2003.pdf>.

example chemical and biological, we reserve the right to review this assurance if the future threat, development and proliferation of these weapons make it necessary".[15]

4.2 The Middle East and Counter-Proliferation

The Middle East became the main testing ground for the Bush doctrine of pre-emption and counter-proliferation. Perhaps the most significant case was that of the 2003 Anglo-American invasion of Iraq, which was widely characterized as the first application of the Bush doctrine. In 2002, the United States and Britain launched a world political and media campaign charging Iraq with possessing WMD in violation of Security Council resolutions, and with having links with al-Qaeda and the 9/11 attacks. As a result, the UN inspection system was reinstalled. Inspectors, however, came up with no evidence to corroborate the Anglo-American charges. Nevertheless, the United States and Britain persisted in their charges and mobilized their forces in the Arabian Gulf in preparation for a possible invasion of Iraq. The Anglo-American military preparations continued on the basis that Iraq represented a serious threat to regional security and stability. These charges were also shown to be false when the Arab countries, with the exception of Kuwait, achieved reconciliation with Iraq in the Beirut Arab summit of March 2002. The Arab–Iraqi reconciliation was made despite mounting US pressure to convince Arab countries to join the Anglo-American campaign against Iraq. The lack of 'credible' evidence of Iraq's possession of WMD as well as Arab conciliation with Iraq did not lead the United States and Britain, however, to reconsider their invasion plans. On 20 March 2003, the two powers launched the Operation Iraqi Freedom campaign to disarm Iraq.

The Anglo-American military campaign against Iraq was unprecedented in the history of modern warfare in terms of the scale of destruction. On the first day of the invasion, US Defence Secretary Donald Rumsfeld stated, "What will follow will not be a repeat of any other conflict. It will be of a force and scope and scale beyond what has been seen before" (cited in Garden 2003: 704). During the invasion, the United States fired more than 380 Tomahawk missiles on Iraq in a single day, and more than 30,000 bombs and 20,000 cruise missiles over a period of 2 weeks. This may be compared to a total of approximately 300 missiles that were fired on Iraq during the 1991 Gulf War (Ismael and Fuller 2007: 450). Indeed, the invasion resulted in the worst human tragedy of the early twenty-first century and the virtual destruction of an entire Arab state. This was followed by a frightening period of chaos and violence, then a long foreign occupation. More than one million Iraqi civilians were killed as a direct outcome of the invasion and subsequent US military operations in

---

15 *Securing Britain in an Age of Uncertainty: The Strategic Defence and Security Review*, October 2010; available at: <http://www.direct.gov.uk/prod_consum_dg/groups/dg_digitalassets/@dg/@en/documents/digitalasset/dg_191634.pdf>.

Iraq.[16] The occupation also resulted in the destruction of Iraqi social capital, which was one of the richest in the Arab world. This took place at two main levels.

The first level is the creation of one of the largest population displacement crises worldwide and the largest in the Middle East since Palestinians were displaced following the creation of the State of Israel in 1948. In 2008, the *UN Office of the High Commissioner for Refugees* (UNHCR 2008) estimated that approximately 4.7 million Iraqis became refugees after the Anglo-American invasion of the country. Out of this figure, an estimated 2 million became refugees in neighbouring countries, representing nearly 10 % of Iraq's pre-war population, and an estimated 2.7 million were internally displaced inside Iraq. The second level is the systematic destruction of the principal purveyors of nationalist consciousness, historical memory and scientific thought in Iraq, represented in the physical elimination of academics, intellectuals, scientists and professionals, and the looting of libraries, museums, census bureaus, schools, cultural centres and medical facilities. In sum, the invasion of Iraq, which was carried out outside the umbrella of the UN Security Council, marked the ultimate extent to which the United States was willing to go in implementing its pre-emption and counter-proliferation strategy in a post-9/11 global order.

Although all US justifications for invading Iraq remained unsupported by disclosed evidence, the Bush administration remained committed to a policy of coercive counter-proliferation in the Middle East. In this context, Syria became the second target of the US counter-proliferation doctrine. Immediately after the termination of major military operations in Iraq in April 2003, President Bush shifted the emphasis to Syria when he declared that "Syria must cooperate… Damascus might have chemical weapons…Every situation requires a different reaction. We are now dealing with Iraq, and afterward we will deal with Syria".[17] In a testimony before the House International Relations Committee concerning Syrian WMD programmes, Bolton (2003) characterized Syria as a 'rogue' state with clandestine nuclear weapons ambitions, and accused it of gaming the IAEA to avoid intrusive inspections of its nuclear-related facilities. In a more aggressive tone, Bolton asserted that "In Syria we see expanding WMD capabilities and continued state sponsorship of terrorism. As the President has said, we cannot allow the world's most dangerous weapons to fall into the hands of the world's most dangerous regimes, and will work tirelessly to ensure this is not the case for Syria".

The US counter-proliferation doctrine was tested on the ground when the United States gave Israel the green light to strike a Syrian industrial facility in the

---

[16] The estimate of more than one million violent deaths in Iraq was confirmed by *Opinion Research Business* (ORB), an independent British polling agency, in September 2007. The ORB poll estimated 1,220,580 violent deaths since the US invasion. In January 2008, ORB published an update based on additional work carried out in rural areas of Iraq. Some 600 additional interviews were undertaken and as a result of this the death estimate was revised to 1,033,000. This is consistent with a 2006 study conducted by the Johns Hopkins University School of Public Health, and published in *The Lancet*. The study estimated 601,000 people killed due to violence as of July 2006. If updated on the basis of deaths since the study, this estimate would also be more than a million.

[17] *Al-Ahram* (Cairo), 14 April 2003.

area of al-Kibar suspected of producing nuclear-related materials. On 6 September 2007, Israel launched 'Operation Orchard' whereby Israeli warplanes violated Syrian airspace, dropped munitions on the suspected target and returned home without experiencing any casualties. Indeed, this was not the first time Israeli warplanes carried out operations against targets inside Syria. In October 2003, Israeli jets attacked a training camp for Palestinian militants near the Syrian capital, Damascus. Israeli warplanes also buzzed the Syrian presidential palace during the summer 2006 war in Lebanon, but did not release any munitions. Following a period of international silence, the United States justified 'Operation Orchard'. In April 2008, the CIA released reports that identified the Syrian target as a nearly completed nuclear reactor secretly under construction since 2001, and provided a 12 min video that made a strong case that the target was a North Korean-built reactor designed for producing weapons-usable plutonium. The strike was launched, however, despite the lack of any imminent or formidable Syrian threat to the security of Israel or the United States. As explained by Spector and Cohen (2008: 6),

> Israel's strike on al-Kibar in September 2007 was, in effect, a clear application of the Bush, internationally disfavoured doctrine. Given that the al-Kibar reactor had not started to operate and, according to the CIA, Syria's fuel fabrication and reprocessing facilities had not been discovered and might not yet have been completed, Syria was unquestionably some time away from producing fissile material for nuclear weapons and still further from producing the weapons themselves. Thus, few could argue that Israel met the traditional necessity/imminence standard in the case of the al-Kibar reactor strike. (The same would be true if the reactor was, in fact, part of an Iranian nuclear weapon programme.) Moreover, Israel did not exhaust or apparently ever initiate other diplomatic means for dealing with this threat.

In the same vein, Iran became a target of the US counter-proliferation doctrine. Following the invasion of Iraq in 2003, the Iranian nuclear programme came to the forefront of the US counter-proliferation agenda in the Middle East and the Arabian Gulf. The United States claimed that Iran was enriching uranium, and that such an enrichment programme would enable that country to go nuclear in the near future. As a result, the United States demanded that Iran must stop its uranium enrichment activities, and commit to a fully accountable civilian nuclear programme, verifiable through the IAEA. The EU Troika (Britain, France, and Germany) took the lead in negotiating with Iran to halt its enrichment programme. When these efforts failed, the United States imposed a series of economic, financial and military sanctions on Iran through three successive UN Security Council resolutions. These were Resolutions 1737, 1747, and 1803, which banned WMD-related trade with Iran, froze the assets of Iran's nuclear and related entities and personalities, prevented Iran from transferring arms outside Iran, banned or required reporting on international travel by named Iranians, called for inspections of some Iranian sea and airborne cargo shipments, and called for restrictions on dealings with some Iranian banks. The United States further escalated its pressure by threatening to resort to military force to stop Iranian nuclear ambitions. This was done despite the fact that the IAEA, which was charged with the task of determining the purpose of Iran's nuclear programme, had not found any conclusive evidence that corroborated American allegations. Beyond several suspicious dual-use items that were found in and around Iran's nuclear facilities, no definitive evidence has yet been found to disprove Iran's use of nuclear energy for solely peaceful purposes.

The US military posture against Iran dates to President Bush's State of the Union address of 29 January 2002, in which he included Iran in the "Axis of Evil" and charged that Iran was a sponsor of terrorism. Bush's remarks gave some indications that the United States might use force against Iran if the latter did not halt its nuclear programme. In 2004, the United States took the campaign to a higher level when Mr. Bolton, in a speech to the US Senate on 28 April, claimed that Iran was involved in dangerous activities that threatened regional stability, which could have adverse repercussions on US and international security. He also made it clear that his government would take all necessary steps, including the use of force if necessary, to protect American interests. In January 2006, President Bush described a nuclear-armed Iran as "a grave threat to the security of the world", language similar to that he had used before the 2003 invasion of Iraq. Two months later, Bush stressed that all options, including the use of military force, were on the table to stop the Iranian nuclear programme. Vice President Cheney also joined the line when he warned of "meaningful consequences" if Iran did not give up its nuclear aspirations.[18]

What was more alarming was the release of reports, presented to the White House by the Pentagon in 2006, on the potential use of US tactical nuclear bombs—'mini-nukes'—to attack Iran's underground nuclear facilities, such as the Natans nuclear reactor which was reported to contain "underground floor space to hold fifty thousand centrifuges, and laboratories and workspaces buried approximately 75 feet beneath the surface" (Hersh 2006). These reports confirmed an earlier classified Pentagon report, released by the *Los Angeles Times* in 2002, that the United States is preparing contingency plans for the possible use of nuclear weapons not only against the three 'axis of evil' nations, but against China, Russia, Libya, and Syria as well. The report stated that nuclear weapons could be used in three types of situations, "against targets able to withstand nonnuclear attack; in retaliation for attack with nuclear, biological or chemical weapons; or in the event of surprising military developments".[19] When specifically questioned about the potential use of nuclear weapons against Iran, President Bush said that "all options were on the table". According to the *Bulletin of the Atomic Scientists*, President Bush "directly threatened Iran with a pre-emptive nuclear strike…It is hard to read his reply in any other way" (Norris and Kristensen 2006: 69). In October 2007, Bush went further by claiming that the issue with Iran is not only the nuclear programme, but also "the knowledge necessary to make a nuclear weapon".[20] This

[18] "U.S. Is Studying Military Strike Options on Iran." *The Washington Post*, 9 April 2006; available at: <http://www.washingtonpost.com/wp-dyn/content/article/2006/04/08/AR2006040801082.html>.

[19] "U.S. Works Up Plan for Using Nuclear Arms." *Los Angeles Times*, 9 March 2002; available at: <http://articles.latimes.com/2002/mar/09/news/mn-31965>.

[20] "Nuclear Armed-Iran Risks World War, Bush Says." *The New York Times*, 18 October 2007; available at: <http://www.nytimes.com/2007/10/18/washington/18prexy.html>.

implied that he wanted Iran to stop all research activities in the nuclear field, which was reminiscent of his emphasis on removing an entire generation of Iraqi nuclear scientists after the invasion of Iraq.

4.3 Counter-Proliferation under Obama

The advent of the Obama administration to power in January 2009 raised hopes that the United States would abandon the Bush counter-proliferation doctrine. In April 2009, President Obama presented an ambitious three-part strategy for non-proliferation and nuclear disarmament. The US strategy included (i) proposing measures to reduce and eventually eliminate existing nuclear arsenals, (ii) strengthening the NPT and halting proliferation of nuclear weapons to additional states, and (iii) preventing terrorists from acquiring nuclear weapons or materials.

In April, 2010, the United States completed a new NPR, which was the first to be published in an unclassified form. The 2010 NPR placed the prevention of nuclear terrorism and proliferation at the top of the US foreign policy and security agenda. It narrowed the role of US nuclear weapons in the overall US defence posture by declaring that "the fundamental role of U.S. nuclear weapons… is to deter nuclear attack on the United States, our allies, and partners". It stated that the United States "would not use or threaten to use nuclear weapons against non-nuclear weapons states that are party to the Nuclear Non-Proliferation Treaty and in compliance with their nuclear non-proliferation obligations". For recognized nuclear powers, such as Russia and China, and states not compliant with the NPT and other non-proliferation obligations, the new NPR made it clear, however, that the United States would reserve the right to use nuclear weapons first or in response to an attack even if that attack did not involve nuclear weapons. This potential use of nuclear weapons, according to the NPR, would only take place in "extreme circumstances to defend the vital interests of the United States or its allies and partners". In addition, the 2010 NPR affirmed that the United States would not develop new nuclear warheads or pursue new military missions or new capabilities for nuclear weapons. This represented a significant change from the 2001 NPR, which emphasized the need for new types of "nuclear warheads that reduce collateral damage" as well as "possible modifications to existing weapons to provide additional yield flexibility". The 2010 NPR also set the stage for additional reductions in US nuclear forces beyond the force levels outlined in the *American–Russian New Strategic Arms Reduction Treaty* (New START), which was signed in April 2010.[21]

On 12–13 April 2010, the United States hosted the Nuclear Security Summit. The Summit brought together 49 states in an effort to foster cooperation and consensus on the question of nuclear disarmament, including the four nuclear weapons NPT member states, three nuclear states not party to the NPT (Indian, Israel,

---

[21] *The Nuclear Posture Review Report*. US Department of Defense, April 2010; available at: <http://www.defense.gov/npr/docs/2010%20nuclear%20posture%20review%20report.pdf>.

and Pakistan), and several non-nuclear states that were NPT parties and members of the Non-Aligned Movement. However, the Obama administration excluded Iran, Syria and North Korea from the guest list. The Summit participants issued a broad communiqué that affirmed their dedication to preventing nuclear terrorism, and agreed on a more specific, but voluntary, Work Plan. The communiqué identified nuclear terrorism as "one of the most challenging threats to international security", and called for the adoption of strong nuclear security measures to "prevent terrorists, criminals, or other unauthorized actors from acquiring nuclear materials". It endorsed President Obama's initiative to secure within 4 years "all vulnerable nuclear material". Towards this end, the participants agreed to:

- Maintain effective security of all nuclear materials, which includes nuclear materials used in nuclear weapons, and nuclear facilities under their control; to prevent non-state actors from obtaining the information or technology required to use such material for malicious purposes;
- Work cooperatively as an international community to advance nuclear security, requesting and providing assistance as necessary;
- Support the objectives of international nuclear security instruments, including the Convention on the Physical Protection of Nuclear Material, as amended, and the International Convention for the Suppression of Acts of Nuclear Terrorism, as essential elements of the global nuclear security architecture;
- Reaffirm the essential role of the International Atomic Energy Agency in the international nuclear security framework and will work to ensure that it continues to have the appropriate structure, resources and expertise needed to carry out its mandated nuclear security activities in accordance with its Statute, relevant General Conference resolutions and its Nuclear Security Plans;
- Recognize the need for cooperation among States to effectively prevent and respond to incidents of illicit nuclear trafficking; and agree to share, subject to respective national laws and procedures, information and expertise through bilateral and multilateral mechanisms in relevant areas such as nuclear detection, forensics, law enforcement, and the development of new technologies; and
- Support the implementation of strong nuclear security practices that will not infringe upon the rights of States to develop and utilize nuclear energy for peaceful purposes and technology and will facilitate international cooperation in the field of nuclear security.[22]

Although the Obama administration claimed the adoption of a regime-oriented view on arms control and non-proliferation, its policies did not deviate dramatically from the Bush doctrine. In fact, the Bush and Obama doctrines share similar assumptions about the ultimate goal of the American non-proliferation strategy in the Middle East, which is to target the Arabs and Iran for disarmament, while leaving Israel's nuclear arsenal intact. For example, the final communiqué of the 2010

[22] See the text of "The Final communiqué of the Washington Nuclear Security Summit," April 2010; available at: <http://fpc.state.gov/documents/organization/140355.pdf>.

nuclear security summit focused on preventing the proliferation of nuclear-weapon material to non-nuclear state and non-state actors, while it did not refer in any way to the disarmament of states which already posses nuclear weapons. This would definitely work to the advantage of Israel as the only nuclear power in the Middle East. Similarly, the 2010 NPR did not deviate significantly from the one articulated in 2001 under the Bush administration. Although the 2010 NPR affirmed that the United States would not use nuclear weapons against non-nuclear weapons states that are party to the NPT and in compliance with their nuclear non-proliferation obligations, it did not exclude the right to use nuclear weapons against states which, from an American perspective, are not compliant with the NPT and other non-proliferation obligations. More importantly, the 2010 NPR affirmed the US right to use nuclear weapons first or in response to an attack even if that attack does not involve nuclear weapons. This reiterates the language of the Bush doctrine with respect to pre-emption and the potential use of nuclear weapons against US adversaries.

In fact, the use or threat of use of military power as one potential tool for disarmament and non-proliferation was evident in the Obama administration's policy toward Iran. Although the administration came to power with a declared commitment to a policy of engagement with Iran over its nuclear activities, it has not ruled out the use of military force as one of the options available to deal with Iran's nuclear programme. On 25 September 2009, President Obama told a news conference at the conclusion of the G-20 summit in Pittsburgh, Pennsylvania that "Iran is on notice that when we meet with them on October 1 they are going to have to come clean and they will have to make a choice". The alternative to giving up their programme, Obama warned, is to "continue down a path that is going to lead to confrontation". Asked about the prospect of using military force to stop Iran from getting the bomb, Obama told reporters that "with respect to the military, I have always said that we do not rule out any options when it comes to US security interests, but I will also re-emphasize that my preferred course of action is to resolve this in a diplomatic fashion. It is up to the Iranians to respond".[23] On 4 March 2011, President Obama reiterated his administration's position when he declared in a speech before the *American–Israel Public Affairs Committee* (AIPAC) that he would not tolerate a nuclear-armed Iran, and that he would act—with military force if necessary—to prevent that from happening.[24]

Other revelations about US intentions to strike Iran were released by the Lexington Institute, a Washington-based think tank with close links to the Pentagon. On 24 August 2012, the Institute issued a report in which it warned that the United States and other NATO states could be swept into a war with Iran over the next few months. "For many of the parties to this crisis, the future looks decidedly bleak, increasing the incentive to choose a course of action which while not necessarily

---

[23] "U.S. and Allies Warn Iran Over Nuclear 'Deception'". *The New York Times*, 25 September 2009; available at: <http://www.nytimes.com/2009/09/26/world/middleeast/26nuke.html?pagewanted=all>.

[24] "Obama Shifts toward Israel on Iran". *The Wall Street Journal*, 5 March 2012; available at: <http://online.wsj.com/article/SB10001424052970203458604577261430761885886.html>.

a good option, is better than doing nothing", the report read. In addition, the report referred to the military build-ups and deployment of forces by the United States and regional actors in the Gulf region as a strong indicator that a military attack on Iran seems "now a better option than any in the future", noting that:

> Saudi Arabia and its Gulf allies have undertaken major military modernization programs involving, among other things, advanced fighters and theater missile defenses…The U.S. is expanding its military presence in the region. Additional mine countermeasure ships and helicopters have been deployed. The *USS Stennis* carrier battle group is enroute to the region. Israel is deploying additional missile defense units, conducting long-range air operations drills and preparing its population for possible war.[25]

In return, when it came to Israel, the United States continued to demonstrate double standards in its policies toward nuclear proliferation and disarmament by turning a blind eye to Israel's nuclear arsenal. This was done despite Israel's persistent refusal to join the NPT, and to bring its nuclear facilities under IAEA supervision. The United States also remained silent after the then Israeli Prime Minister, Olmert, admitted in 2006 that Israel had nuclear weapons, and that Iran's manoeuvering was intended to acquire the same capacity. In fact, as Iran continued to be subjected to sanctions for its nuclear development activities, the United States and Israel concluded a nuclear agreement in April 2008 to upgrade safety and technology at the Israeli nuclear reactor of Dimona. According to *Hareetz*, the agreement would "enable the Israel Atomic Energy Commission to access most of the latest nuclear safety data, procedures and technology available in the U.S."[26] On 7 July 2010, Israel's Army Radio reported that the United States pledged to sell Israel nuclear technology and other supplies for nuclear energy production, despite the fact that Israel is not a signatory of the NPT. According to Army Radio's diplomatic correspondent, the US offer would "put Israel on a par with India, another NPT holdout which is openly nuclear-armed but in 2008 secured a U.S.-led deal granting it civilian nuclear imports".[27]

In September 2009, the United States voted against two IAEA resolutions which called for the establishment of a nuclear-weapons-free zone in the Middle East. The first resolution, sponsored by Egypt, called upon "all states of the region, pending the establishment of the zone, not to develop, produce, test or otherwise acquire nuclear weapons or permit the stationing on their territories or on territories under their control of nuclear weapons". The second resolution, issued at the end of the fifty-third annual general assembly of the IAEA and sponsored by a group of Arab states, called upon Israel to put all its nuclear sites under UN inspection, and sign the NPT. Prior to the vote, Davies, the US representative to the IAEA, described the resolution as 'redundant', warning that "such an approach

---

[25] "The Guns Of October—In The Persian Gulf." *Lexington Institute*, 24 August 2012; available at: <http://www.lexingtoninstitute.org/the-guns-of-october--in-the-persian-gulf?a=1&c=1171>.

[26] "Israel and U.S. sign nuclear cooperation agreement." *Hareetz*, 14 April 2008; available at: <http://www.haaretz.com/news/israel-and-u-s-sign-nuclear-cooperation-agreement-1.243947>.

[27] "Report: Secret document affirms U.S.-Israel nuclear partnership."*Haaretz*, 7 July 2010; available at:<http://www.haaretz.com/news/diplomacy-defense/report-secret-document-affirms-u-s-israel-nuclear-partnership-1.300554 >.

is highly politicized and does not address the complexities at play regarding crucial nuclear-related issues in the Middle East".[28] In a press conference with Israeli Prime Minister Netanyahu on 6 July 2010, President Obama reiterated US unconditional support for Israel's nuclear policy. "There is no change in U.S. policy when it comes to [nuclear non-proliferation]...[Israel has] got to be able to respond to threats or any combination of threats in the region...We remain unwavering in our commitment to Israel's security. And the United States will never ask Israel to take any steps that would undermine their security interests", Obama said.[29] These policy stances, among others, are a clear indication of the prejudiced American approach to counter-proliferation in the Middle East.

References

Bolton, J. (2003). Syria's weapons of mass destruction and missile development program. A Testimony before the House International Relations Committee, Washington DC. Retrieved September 13, 2003, from http://merln.ndu.edu/archivepdf/syria/State/24135.pdf.

Dokos, T. P. (2008). *Countering the proliferation of weapons of mass destruction: NATO and EU options in the Mediterranean and the Middle East.* London: Routledge.

Garden, T. (2003). Iraq: The military campaign. *International Affairs*, *79*, 4.

Hersh, S. (2006, April 17). The Iran plans: Would President Bush go to war to stop Tehran from getting the bomb? *The New Yorker* (Annals of National Security).

Ismael, T. Y., & Fuller, M. (2007). The disintegration of Iraq: The manufacturing and politicization of sectarianism. *International Journal of Contemporary Iraqi Studies, 2*, 3.

Khurana, G. (2004, April–June). Proliferation security initiative: An assessment. *Strategic Analysis, 28*, 2.

Litwak, R. (2007). *Regime change: U.S. strategy through the prism of 9/11.* Baltimore, MD: Johns Hopkins University Press.

Newman, A. (2004). Arms control, proliferation and terrorism: The Bush Administration's Post-September 11 Security Strategy. *The Journal of Strategic Studies*, *27*, 1 (March).

Norris, R. S., & Kristensen, H. M. (2006, September/October). U.S. nuclear threats: Then and now. *Bulletin of the Atomic Scientists*, *62*, 5.

Record, J. (2004, July 8). Nuclear deterrence, preventive war, and counter proliferation. *Policy Analysis*, *519*.

Schmitt, G. (2001). Memorandum to opinion leaders. Project for a New American Century, 13 December.

Spector, L., & Cohen, A. (2008, July/August). Israel's airstrike on syria's reactor: Implications for the nonproliferation regime. *Arms Control Today.*

UNHCR. (2008). Iraq: Latest return survey shows few intending to go home soon. Briefing Note, 29 April.

Yost, D. S. (2006). France's new nuclear doctrine. *International Affairs*, *82*, 4.

[28] "IAEA passes nuke-free Middle East resolution". *Al-Arabiya*, 18 September 2009; available at: <http://www.jpost.com/International/UN-criticizes-Israels-atomic-program>; "UN criticizes Israel's atomic program". *The Jerusalem Post*, 18 September, 2009; available at: <http://www.jpost.com/servlet/Satellite?cid=1253198154270&pagename=JPost%2FJPArticle%2FShowFull>.

[29] "Remarks by President Obama and Prime Minister Netanyahu of Israel in Joint Press Availability". The White House, Office of the Press Secretary, 6 July 2010.

Chapter 5
Arab Perceptions of Global Arms Control Approaches

Scholars of international politics have long recognized that human behaviour is largely shaped by how reality is perceived and evaluated, and that comprehending decision-makers' cognition of reality is crucial for understanding their behaviour (Jervis 1976). In fact, the cognitive approach to international politics is based on these premises. The difference between various cognitive schools lies in their identification of the locus of the most crucial cognitive variables, such as perceptions, beliefs, images, and values. In the meantime, they all share the assumption that national leaders make decisions within the constraints of 'bounded rationality'. These constraints are related to the external situation as well as the capacities of the decision-maker. In this respect, one can distinguish between (i) external boundaries, which include missing, erroneous, or unknowable information about external crises, and (ii) internal boundaries to rational decision-making, which are the result of policymakers' limited information processing capacity when studying exceptionally complex issues. Instead of searching all information for the best outcome, policymakers usually select an alternative that is acceptable and compatible with their existing views (Mintz and DeRouen 2010: 68–69).

In evaluating Arab perceptions of arms control, it is hardly possible to determine a unified Arab perception. In fact, one can differentiate between two dimensions. The first is the distinction between the perceptions of different Arab countries, and the second is the distinction between elites and mass perceptions.

In the first dimension, although Arab countries share the two major commonalities of Arabism and Islamism, they differ in their security perceptions and commitments, and sometimes in their foreign policy orientations. One can distinguish between the perceptions of three categories of Arab countries: (i) Arab Maghreb countries (Algeria, Libya, Morocco, Mauritania, and Tunisia); (ii) Arab Mashreq countries (Egypt, Iraq, Jordan, Palestine, Lebanon, and Syria); and (iii) Yemen and *Gulf Cooperation Council* (GCC) Arab states, namely Bahrain, Kuwait, Qatar, Oman, Saudi Arabia, and the United Arab Emirates (UAE). Perceptions of security issues in general, and arms control in particular, vary as we move from one Arab sub-region to the other. This is because the distances between each Arab country

G. M. Selim, *Global and Regional Approaches to Arms Control in the Middle East*, SpringerBriefs in Environment, Security, Development and Peace 4,
DOI: 10.1007/978-3-642-29314-6_5, © The Author(s) 2013

and each issue are not the same. Whereas security-related issues, including arms control and non-proliferation, dominate in the Arab Mashreq and GCC states, they are of less importance in the Arab Maghreb.

On the second level, there are wide variations between the perceptions of the ruling elites and those of civil society on the question of arms control. At the level of the ruling elites, there is a gap between actual and perceived threats. Whereas the American counter-proliferation approach after 9/11 represented a threat to Arab national security, ruling elites in Egypt (under the Mubarak regime), Jordan, and the GCC states did not perceive them as such. On the contrary, they endorsed the American project and facilitated its implementation. This can be accounted for by the American 'existential' pressures on Arab regimes to cooperate with the American design in the region, or face assured collapse. In this respect, Arab ruling elites acted to secure regime survival, as regime collapse was viewed as the most crucial security threat. This was obvious in the case of Libya under Qaddafi, which abandoned its nuclear programmes under American threats. This was done despite Libya's continued emphasis on the Israeli nuclear arsenal as a security threat. In this case, regime survival was perceived as more important than deterring Israel.

At the level of civil society, one can argue that Arab civil society is generally more radical than the ruling elites on security-related questions. Civil society radicalism is reflected in a number of issues. For example, Arab civil society, unlike the ruling elites, was very critical of the US invasion of Iraq, and called for the defeat of the American project in Iraq. Arab civil society is also quite critical of any normalization with Israel before the full withdrawal of Israeli forces from Palestinian, Syrian, and Lebanese territories. Contrary to the dominant views of the ruling elites, it considers Israel as the main security threat to Arab national security. This disproves the widely held view that civil society interactions across the region are one of the main strategies for promoting peace.

Because perceptions of the ruling elites are the perceptions that matter in the areas of decision-making, resource mobilization, and foreign policy commitment, we will review their perceptions of the questions of arms control and non-proliferation, and attempt to contrast them with mass and civil society perceptions in the Arab Mashreq and GCC countries. Our emphasis on these two sub-regions is justified on the basis that (i) they have been the focus of global arms control projects in the Middle East, and (ii) the regional arms control debate has been mostly articulated by the Arab Mashreq and GCC countries, in addition to non-Arab actors, mainly Israel and Iran.

In surveying how arms control is perceived by the Arab Mashreq and GCC countries, one encounters two major research problems. The first relates to the sources from which one can extract the security perceptions of the ruling elites. This area is rarely a matter of public discourse in Arab countries. In addition, Arab countries lack a policy of public access to declassified state documents. For example, it is quite difficult to access the verbatim record of the various proposals submitted by Arab countries in the deliberations of the Middle Eastern multilateral working groups or the NATO—Mediterranean Dialogue. One has to rely upon the verbal articulations of governmental officials to know the content of such proposals.

The second problem relates to the divergence between declared and real perceptions, and between perceptions and policies. In some cases, declared articulations on the questions of security and arms control are directed toward gaining domestic legitimacy or countering domestic opposition, but these articulations are not injected into the policy-making process. For example, despite Egypt's public criticism of Israel's atrocities in the Palestinian occupied territories and Lebanon, Egyptian–Israeli trade ties have increased. This calls for a systematic search for the real perceptions, and matching perceptions with actual policies with a view to locating areas of incongruence. It also calls for shifting the analysis from the level of perceptions to the higher level of 'conceptualizing' the issues at stake. Conceptualization entails not only the awareness of certain issues and the policy stances adopted, but also the rationale provided by the actors to justify and account for such policy stances. However, conceptualizations and perceptions cannot be easily separated. Although they are analytically different, they are interconnected in reality. As a result, both categories will be used interchangeably.

If one reviews the Arab discourse on arms control over the last two decades, one can conclude that Arab perceptions have undergone a drastic change. In this context, one can distinguish between two main stages of Arab perceptions since the rise of the arms control debate in the region in the early 1990s. The first stage begins with the revival of the arms control agenda following the end of the Cold War and continues up to the American invasion of Iraq in 2003. The second stage extends from 2003 up to the outbreak of the Arab Spring uprisings from 2010 and onward. A detailed review of these perceptions is in order.

5.1 First Stage: Arab Emphasis on Comprehensive Arms Control

During the 1990s, the Arabs conceptualized arms control as an important instrument for the durable resolution of the Arab–Israeli conflict. They were quite attentive to the centrality of that strategy given their perception of the regional imbalance in favour of Israel. In their assessment, a Middle Eastern arms control regime should deal with conventional and non-conventional weapons in the region. It should also focus on the limitation of all levels of armaments, and the banning of all WMD on a comprehensive and equitable basis. The Middle East must be declared a zone free from weapons of mass destruction. This meant that Israel had to sign and ratify the NPT, and that the Arabs would link the signing of the CWC to Israel's adherence to the NPT.

Arab conceptualizations of arms control were articulated within the framework of two regional projects. The first was the Egyptian Initiative on the Ban of Weapons of Mass Destruction, which was announced by former President

Mubarak in April 1990 and formally presented in a letter to the UN Secretary-General in July 1991. The Initiative reflected Egypt's growing concerns with Israel's nuclear programme and its impact on the proliferation of WMD programmes in the region, especially after the revelations about Iraqi capabilities in the planning and production of WMD. The Initiative represented an expansion of an earlier joint Egyptian–Iranian proposal presented in 1974 at the UN General Assembly for the establishment of a nuclear-weapons-free zone in the Middle East. The 1990 Initiative, however, was more comprehensive as it covered all types of WMD. It emphasized that all WMD (nuclear, chemical, and biological) without exception in the Middle East should be prohibited by calling upon all states in the region to join the NPT and adopt IAEA safeguards on all nuclear facilities. It also stipulated that all states of the region without exception should make equal and reciprocal commitments in this regard. Further, the Initiative called for the establishment of verification measures and modalities in order to ascertain full compliance by all states in the region with the full scope of that prohibition without exception (Karem 1995: 130).

The Egyptian Initiative was later expanded in a paper delivered by former Egyptian foreign minister Amr Moussa to the Conference on Disarmament in Geneva in July 1991. The updated version of the Initiative called upon the states of the region to endorse the *Weapons of Mass Destruction Free Zone* (WMDFZ) in declarations to the UN Security Council and to confirm their intention to refrain from actions that would impede the establishment of such a zone. It also called upon all regional actors to declare their readiness not to use nuclear, chemical or biological weapons; produce or acquire nuclear weapons; or produce or acquire nuclear weapons material. Further, the Initiative called for an active UN involvement in the verification of arms control and disarmament measures in the region.[1]

In September 1992, the Arab League adopted a draft resolution in support of the Egyptian Initiative. The resolution (no. 5232) called for freeing the Middle East from all types of WMD as the best guarantee for achieving security and stability in the region. It also called for the creation of a follow-up committee, consisting of the Arab states participating in the Conference on Disarmament, in order to coordinate the Arab position with other international forums and groups.[2] In the same vein, Syria and Saudi Arabia endorsed the Egyptian Initiative. In December 1994, the two countries joined Egypt in issuing a communiqué in support of the Initiative. The tripartite communiqué called on the international community, particularly the co-sponsors of the peace process, to work seriously towards the realization of the WMDFZ in the Middle East and to put pressure on Israel to adhere to this objective.[3]

Whereas Israel accepted in principle the idea of establishing a WMDFZ in the Middle East, it differed with Egypt on the necessary mechanism for its

---

[1] See Conference on Disarmament Doc. CD/1098, 21 July 1991.

[2] See the text of Arab League resolution 5232 for Session No. 98, 13 September 1992.

[3] See text of the tripartite communiqué, cited in Feldman 1997, p. 229.

implementation. On the one hand, Egypt viewed the establishment of a WMDFZ as a prelude to or part of an overall peace settlement in the region. On the other hand, Israel contended that the objective of the WMDFZ could only be realized after the establishment of peace and the resolution of all unsettled issues between the Arabs and Israel. Further, Israel did not welcome an active role for the UN in the verification area and called instead for the establishment of an independent regional institution that would be responsible for verification tasks (Pande 1998: 5).

The second project was the ACRS working group. In the ACRS discussions, the Arab states contended that the arms control negotiations should focus on the limitation of all levels of armaments and a ban on all types of WMD. The Arabs gave priority to the Israeli nuclear weapons and the need to control the arms race and to eliminate WMD from the Middle East on a comprehensive and equitable basis. They contended that the Israeli nuclear programme represented the most serious threat to the security and stability of the region. In addition, Israel's nuclear programme, the Arabs argued, would trigger the development of other WMD programmes in the region. In this respect, the Arabs stipulated that Israel must sign the NPT and accept the IAEA safeguards on its nuclear facilities as a precondition for a comprehensive peace settlement in the region. They proposed that, at the least, Israel should accept a specific date or set of conditions at which time it would renounce its nuclear ambiguity and join the NPT. Further, they linked the endorsement of the Additional Protocol of the NPT to Israel's endorsement of the NPT itself. This went counter to the Israeli approach, which advocated the introduction of CBMs as a prelude for any discussion on structural arms control and WMD. Once confidence was established and peace was concluded between the regional parties, Israel argued, the more difficult issues relating to WMD and structural arms control could be addressed (For details on Arab and Israeli approaches within the ACRS, see Jones 2003: 137–138; Steinberg 1994: 127–135; Said 1995: 29–37). Indeed, the United States fully supported the Israeli position in the ACRS negotiations. On 28 January 1992, the then US Secretary of State James Baker made it clear that the agenda of the ACRS talks should focus on "a set of modest confidence-building or transparency measures covering notifications of selected military-related activities and crisis-prevention communications". The purpose, Baker added, would be "to lessen the prospects for incidents and miscalculations that could lead to heightened competition or even conflict".[4]

Arabs were also quite critical of the global arms control proposals in the Middle East and viewed them as strategies for perpetuating Israel's military advantage over the Arab states. The Arab rejection of the global arms control proposals were based on the following grounds.

[4] Remarks by Secretary of State James A. Baker before the Organizational Meeting for Multilateral Negotiations on the Middle East, held at the House of Unions, Moscow, 28 January 1992; available at: http://www.mfa.gov.il/MFA/Archive/Peace%20Process/1992/SECRETARY%20OF%20STATE%20BAKER-%20ORGANIZATIONAL%20MEETING%20F.

First was the lack of consultation with Arab countries on the formulation of the arms control projects. Virtually all external projects for arms control in the region were unilaterally designed by Western sponsors. Arab states were viewed as a domain for implementation, rather than as partners in the formulation and implementation of an arms control agenda. They were pressured to accept the global arms control projects and deal with them as if they were facts. This is reminiscent of the old Middle East project of the 1950s which viewed the region as a mere geographical expression and an arena for Cold War competition. Some Arab countries were invited into the NATO—Mediterranean Dialogue and the EMP, but only as 'legitimizers' of the Western plans, rather than as partners in the process.

Second, the global arms control proposals were vague in many important aspects. The US Plan, for example, did not define the meaning and boundaries of the term 'destabilizing arms'. Rather, it left the issue subject to the assessment of the permanent members of the UN Security Council (Al-Dessouki 1996: 11–12). The multilateral proposals were also vague about the criteria for arms sales. These criteria did not mean much in reality as they were left to the judgement of the arms exporting countries. When the permanent members met in Washington in May 1992, they failed to reach an agreement on the operational criteria to be applied for arms exports. From an Arab perspective, the main problem with such an ambiguity was that it could result in arms sales policies that would be substantially influenced by the political interests of the arms suppliers. According to Abdel-Salam (1992: 245), an Egyptian analyst, the ambiguity of the arms sales criteria would have a negative impact on the Arab states. Given the record of Western arms sales policies in the region over the last few decades, the proposed criteria, he argued, would be employed to deny the Arab states access to advanced weaponry systems. He explained that:

> Whereas the great powers often view any possible use of weapons by Israel as a defensive use, even if such weapons are used within a framework of a clear offensive strategy, they understand the use of weapons by Arabs as an offensive and destabilizing act. Meanwhile, there are clauses that become contradictory if applied to one state. For example, Syria is entitled to participate in the regional security arrangements, while at the same time is classified by the West as a terrorism-sponsoring state. The result is that any state in the region, with the exception of Israel, could be denied access to conventional weapons with varying degrees.

Third, the global arms control proposals were based on a strategy of selective application. This selectivity could be noticed in the area of missile and WMD proliferation. In the field of missiles, the US Plan provided the countries which already possessed ballistic missiles with an advantage because it did not specify a timetable for their removal from the region. It also did not provide any verification mechanism for the peaceful use of missile technology. In the same vein, the International Communiqué on Arms Transfers and Non-proliferation focused on freezing the numbers of ballistic missiles in the region without suggesting specific measures or timetables for their removal. This meant that actors which possessed advanced missile capabilities would have an advantage over those with less advanced arsenals. From an Arab perspective, the proposed freeze would be in favour of Israel, given its possession of the most sophisticated missile arsenal in the region.

The same philosophy was also evident in the field of WMD. The US Plan sought to provide Israel with an advantage by freezing the nuclear status quo, characterized by an Israeli nuclear monopoly, and by providing neither a mechanism for the removal of already existing nuclear weapons nor a timetable for the creation of a nuclear-weapons-free zone. In the meantime, the Plan focused on the elimination of chemical and biological weapons in the region. This meant that the Arab states would be obliged to abandon their chemical and biological weapons as their only non-conventional deterrent, while leaving Israel as the only nuclear power in the Middle East for an unlimited period of time. Similarly, through its emphasis on banning weapons-usable nuclear materials and on placing all nuclear activities in the region under IAEA safeguards, the International Communiqué of July 1991 sought to perpetuate Israel's nuclear monopoly in the region. This is because the supervision of the IAEA did not necessarily mean the removal of Israel's nuclear stockpile, while the ban on nuclear material would only affect the Arabs if they tried to develop nuclear weapons programmes. (Al-Dessouki 1996: 13; Diyab 1995: 31).

The selectivity of the global arms control proposals was also noticeable in the exclusion of certain regional actors from the arms control negotiations. The absence of some actors helped undermine the credibility of the arms control proposals as a suggested framework for the establishment of an arms control regime in the region. This was mainly due to the presence of different threat perceptions among the main regional actors. For example, the US Plan excluded Turkey from its definition of the Middle East, which meant that Turkey would have no obligations under the proposed arms control regime. Given the history of political and military crises between Turkey on the one hand and both Syria and Iraq on the other, it came as no surprise that Syria and Iraq rejected the Plan's definition of the region.

Also, the ACRS negotiations saw the absence of key actors in the region such as Syria and Lebanon, who boycotted the ACRS meetings from the beginning, and Iraq, Iran and Libya, who were not invited to participate. It was difficult to imagine how a discussion of a regional WMDFZ could have succeeded in the absence of Iran, Iraq, Libya and Syria, all states suspected at the time of WMD activities. In fact, Israel used the absence of these states to further justify its reluctance to discuss structural arms control arrangements within the ACRS framework. From an Israeli perspective, there were no circumstances under which Israel could have agreed on arms limitations with some Arab actors while the military forces of its major Arab enemies were kept intact. The exclusion of Iraq and Iran from the ACRS negotiations was also a concern to the GCC countries, given the experiences of the Iraq–Iran War and the Iraqi invasion of Kuwait (Jones 2003: 146). In addition, the NATO—Mediterranean Dialogue excluded key Arab Mediterranean countries, such as Libya, Syria, and Lebanon. This exclusion had a negative impact on the overall Arab perception of the Dialogue and its proposed CBMs. In fact, Egypt and other Arab Mediterranean countries viewed the Dialogue with suspicion, and considered it as an American attempt to merge Israel into the Middle East security structure before reaching a comprehensive settlement of the Arab–Israeli conflict (Mekheimar 2002: 46).

Another area of selectivity was evident in the gap between promises and policies. In the post-Cold-War era, CBMs have been proposed as mechanisms for conflict resolution and peace building in Middle Eastern conflicts. However, the experience of the application of CBMs in the Middle East shows that there are limitations to this proposal. Whereas Western powers advocated the pursuit of a CBMs strategy in resolving the Arab–Israeli conflict, they nevertheless refrained from pursuing a similar strategy in other regional conflicts, such as the conflict with Iraq over its invasion of Kuwait in 1990 and implementation of Security Council resolutions, and the conflict with Libya in 1992 over the Lockerbie crisis. In the first case, no CBMs were suggested to deal with the problem of the Iraqi occupation of Kuwait. Instead, military force and severe economic sanctions were used not only to dislodge Iraq from Kuwait but also to force it to comply with UN Security Council resolutions. In the second case, Libya was excluded from the EMP, severe economic sanctions were applied, and no compromise was accepted until the Libyan economy was badly hurt. In both cases, the question of the application of CBMs was never envisaged, and Western powers resorted to a 'compliance' strategy rather than a CBMs strategy.

In addition, the same Western powers which proposed arms control measures in the Middle East have adopted national defence doctrines that undermined their credibility as sponsors of arms control and non-proliferation efforts in the region. These powers dealt a major blow to the NPT, which stipulated that 'legitimate' nuclear powers would eventually remove their nuclear weapons and that the solution to nuclear proliferation was complete nuclear disarmament. Instead of reducing the role of nuclear weapons in their defence doctrines, thus demonstrating a positive model of non-proliferation and disarmament to other countries, Western powers moved in the direction of assigning a more significant role to nuclear weapons in their national security strategies. They also expanded their nuclear doctrines to authorize the use of nuclear weapons in developing countries.

In 1990, NATO adopted the London Declaration, which asserted that NATO member states would keep nuclear weapons indefinitely while new threats emerged which would require their retention. The Declaration emphasized that nuclear weapons would "continue to fulfill an essential role in the overall strategy of the Alliance to prevent war by ensuring that there are no circumstances in which nuclear retaliation in response to military action may be discounted".[5] The 1995 edition of the *NATO Handbook* also affirmed that "the maintenance of an appropriate mix of nuclear and conventional forces based in Europe will be required for the foreseeable future". It went on to argue that the fundamental purpose of NATO's nuclear power is "to preserve peace and prevent war or any kind of coercion".[6]

In March 1990, the US Joint Chiefs of Staff put out their annual Military Net Assessment, which, for the first time, pointed to "increasingly capable Third World threats" as a justification for maintaining US nuclear weapons, and endorsed new

---

[5] See the "London Declaration", *Survival,* September–October 1990, pp. 469–472.

[6] *NATO Handbook* (Brussels: NATO Office of Information and Press, October 1995), pp. 41–42.

roles for nuclear forces, namely to confront Third World countries with WMD capabilities.[7] In March 1991, the Joint Chiefs' Military Net Assessment specifically identified non-strategic nuclear weapons as a class of weapons that "could assume a broader role globally in response to the proliferation of nuclear capability among Third World nations".[8] When the Clinton administration took office in 1993, it initiated a major review of the US defence doctrine to meet the new threats of the post-Cold War era. In April 1993, the US government officially endorsed the expansion of nuclear strategy as military doctrine when the Joint Chiefs issued *The Doctrine for Joint Nuclear Operations*. The document affirmed that the purpose of US nuclear weapons was to deter the use of WMD, particularly nuclear weapons. It also advocated the development of low-yield precision-guided nuclear weapons for possible retaliatory use in regional wars to "avoid destabilizing the conflict". In October 1993, the government issued the 'Bottom-Up Review' document, which authorized the "maintenance of flexible and robust nuclear and conventional forces to deter WMD attacks through the credible threat of devastating retaliation".[9]

Similarly, France expanded its nuclear doctrine. The new French doctrine shifted from the previous principle of deterrence towards an active policy of nuclear intimidation, which implied the possibility of the pre-emptive use of nuclear weapons. In June 1991, the then prime minister Jacques Chirac told the Academy of Moral and Political Science that proliferation of nuclear weapons to the Middle East region meant that Europe and France would have to "radically review their nuclear means and strategies".[10] In June 1992, Chirac (1992) strongly endorsed expanding France's nuclear doctrine to counter the proliferation of nuclear weapons. In February 1994, the French government issued a new Defence White Paper that broadly defined the objective of French nuclear forces. The White Paper, the first one since 1972, stated that "the Cold War is over, but the nuclear era goes on". It explained that "the scenarios in which [nuclear deterrence] may possibly be exercised are diversifying [to include] dealings with existing or new major powers, [and] dealings with regional powers that would threaten our vital interests".[11] In 1995, France resumed nuclear testing in the South Pacific.

No doubt, the expansion of Western nuclear doctrines represented a setback to global arms control proposals and efforts to create a WMDFZ in the Middle East. As explained by Kristensen and Handler (1995) "Prescribing nuclear weapons to counter proliferators muddles the non-proliferation message.... [It] risks strengthening non-nuclear countries' incentives to pursue nuclear weapons and other weapons of mass destruction. The more proliferation becomes intertwined

7 Joint Chiefs of Staff, "1990 Joint Military Net Assessment" (Washington, DC, March 1990).

8 Joint Chiefs of Staff, "1991 Joint Military Assessment" (Washington, DC, March 1991).

9 Les Aspin, US Secretary of Defense, "Report on the Bottom-Up Review". Washington DC, October 1993.

10 "Mr. Chirac: European Defence Must Be Created". *Le Monde*, 19 June 1991.

11 *Livre Blanc sur la Défense, 1994* (Paris: Service d'Information et de Relations Publiques des Armées, Ministère de la Défense, 1994), English excerpts.

with nuclear deterrence, the less likely it is that it will be forestalled." Al-Baradei, former IAEA Director-General, also warned that the West's double-standards approach to nuclear non-proliferation would seriously undermine the NPT regime and global security. He explained:

> Until we see serious movement towards nuclear disarmament by the nuclear weapon states and a security system that does not rely on nuclear deterrence, there will continue to be a sense of double standards…A world of nuclear 'haves and have-nots' is not sustainable. Eventually, some countries—particularly those in areas of conflict—will start to ask whether they are better off leaving the NPT. This would be a terrible development, because a world with more nuclear-weapon states is a much more dangerous world. What we need is to move away from nuclear weapons and not to increase the number of those who have them. This could be the beginning of the end of our world.[12]

Further, the credibility of the global arms control proposals was undermined in another important way. These proposals did not affect Western arms sales in the region. Rather, Western powers continued to support their regional allies with advanced weaponry systems, which further restricted arms control efforts in the region. Between August 1990 and December 1991, the United States announced arms sales totalling over US$19 billion to a number of countries in the region, out of which US$6 billion worth were committed after the announcement of the Bush arms control plan in May 1991. In 1992, additional US arms sales to the region raised this figure to US$24 billion, the largest proportion of which went to Israel, followed by Saudi Arabia and Egypt. Although the US administration argued that such arms transfers were for defensive purposes, the types of weapons covered by these transfers were classified in the 1990 CFE treaty as designed for "launching surprise attack and for initiating large scale offensive action" (Navias 1994: 100).

This upsurge in US arms sales to the region led other major arms suppliers to follow US footprints in the rush to sell arms there, as they became "disillusioned with the USA for advocating restraints on sales to the Middle East whilst making major transfers itself" (Spear 1994: 99). In 1993, Russia marketed approximately 370 weaponry systems through its active participation in the 'Idex 93' arms fair at Abu Dhabi, UAE. The same year, Russia delivered three submarines to Iran. In 1994, the Russian government announced arms sales of US$1.6 billion to Syria, including advanced systems such as the SU-27 and MiG-29 fighters and the up-to-date T-80 tanks. The same year, the Russians also entered negotiations with the Turks for the sale of SS-21 ballistic missiles to Turkey (Klein 1995: 42–44). In 1991, Germany and Israel signed an arms deal by which Israel would get three German-made 'dolphin' submarines capable of launching nuclear missiles. The submarines were delivered to Israel in 1999/2000. In 1992, Britain supplied Saudi Arabia with Tornado fighters as part of an arms deal worth US$17 billion, the biggest-ever arms deal between the two countries. These arms transfers, among others, were contradictory to the logic and promises of arms control, and led accordingly to regional disillusionment with the Western arms control proposals

[12] "Nuclear dynamics," *Al-Ahram Weekly* (Cairo), Issue No. 789, 6–12 April 2006.

and to a charge of hypocrisy. In fact, the prospects for a regional arms control regime faded completely in the face of the desperate competition between major arms supplies in the Middle Eastern market.

Another major obstacle to the establishment of an arms control regime in the Middle East is the presence of a strategic disequilibrium in the region. In the area of non-conventional weapons, Israel is in a superior strategic position relative to all other Arab countries and Iran combined. Israel is the only state in the region that possesses nuclear capabilities, in addition to an efficient system of nuclear delivery vehicles, with no counterparts available to any of its neighbours. Although some Arab actors do possess chemical and biological weapons, they do not possess the capability to deter a first Israeli nuclear strike or launch a deadly first or second strike. Building on its nuclear power, in addition to its satellite reconnaissance assets, Israel is in a position to launch a deadly nuclear strike against its Arab adversaries, yet remain largely immune from an Arab counter-attack.[13] This strategic imbalance is also extended to the area of conventional weapons. Whereas the Arab countries enjoy military superiority from a quantitative perspective, Israel enjoys a qualitative military superiority over the Arabs and Iran thanks to its advanced technological base and Western support.[14]

This strategic imbalance has created serious obstacles for arms control efforts in the region. On the one hand, it is extremely difficult to remove or control the proliferation of WMD under conditions of strategic disequilibrium. States possessing WMD do not feel obliged to give up their weapons, simply because this will reduce their strategic superiority in an international system characterized by anarchy and security dilemmas. Meanwhile, perceptions of security threats will always motivate non-WMD powers to strive to possess such weapons in order to reach a

[13] Contrary to the widely-held belief that nuclear weapons serve only as a deterrent and as a weapon of last resort, they can actually be employed in situations that fall short of a total war waged for absolute annihilation or survival. In fact, nuclear weapons lend themselves readily to tactical use, especially if the attacker perceives no deterrent. The accessibility and tacticization of nuclear weapons have increased the temptation to develop, deploy, and use or threaten to use them in conflict situations, especially when they are perceived as an ultimate security insurance given their ability to inflict tremendous losses, including wiping out entire communities. Israel actually used its nuclear arsenal during the October 1973 War as a compellent. This occurred when nuclear-tipped missiles were put on alert and deployed in their firing position to compel the United States to accelerate its aerial re-supply effort, although the war was being fought on Arab territory and the survival of Israel was not at stake. Later on, Yitzhak Mordechai, the former Defence Minister of Israel, suggested that Israel "had tactical nuclear weapons and would be prepared to use them".

[14] In the assessment of quantitative and qualitative advantages, most Israeli analysts argue that the Arab quantitative advantage counterbalances Israel's qualitative one, thus creating a sort of conventional parity between the two sides. However, this assessment has serious limitations for two reasons. First, it is based on the theoretical assumption that all the Arab countries could be united in a military confrontation against Israel—a highly unrealistic scenario given present patterns of regional and international alliances. Second, the experiences of the 1991 Gulf War and the 2003 Anglo-American invasion of Iraq demonstrated the fact that military technology, rather than the number of troops, is the most decisive factor in determining the outcome of war.

state of equilibrium that could enable them to engage positively with others. The task of persuading the haves to give up what they have and the have-nots not to pursue the path of the haves is certainly not an easy one to achieve. In this respect, the Middle East's strategic disequilibrium has put Israel under no pressure to seek arms control arrangements that could limit its military superiority or oblige it to give up its nuclear armament. This explains Israel's reluctance to achieve any real progress in the ACRS arms control negotiations.

On the other hand, the strategic imbalance has undermined the global approach to arms control in the Middle East which has focused on extending the global regime of freezing the level of armaments in the region. Although this approach was functional within the context of East–West relations where the actors reached a level of military equilibrium, it has been highly problematic in the Middle Eastern context as it meant in practice the 'freezing' of the present military disequilibrium in the region. More importantly, the power imbalance gave the Arabs no leverage over Israel, leaving them with little influence to push for genuine arms control negotiations (Al-Dessouki 1996: 18–19). In this context, the Arabs rejected the military situation that the global arms control proposals tried to legitimize, which was the perpetuation of the strategic disequilibrium in favour of Israel.

As explained by one Egyptian scholar, there is no historical precedent for the operationalization of arms control under conditions of military disequilibrium. Historically, the question of WMD was addressed under one of three conditions; military defeat, strategic collapse, or strategic equilibrium:

> The Iraqi arsenal of WMD was only removed after the military defeat of Iraq in the second Gulf war in 1991. Iraq was forced to accept the unilateral removal of its WMD under the threat of military action including missile attacks by the Americans. The second historical precedent occurred when Kazakhstan, the Ukraine, and Belarus, former Soviet republics, agreed to remove their nuclear weapons after their independence as a result of the collapse of the Soviet Union in 1991…Rooted in a strategic collapse, the nuclear disarmament of South Africa was more or less similar to the experience of the former Soviet republics. The collapsing apartheid regime was coerced to de-nuclearize before the black majority take-over. The third historical model in which WMD were removed was the American–Soviet agreements during the Cold War to reduce nuclear capabilities. Such agreements were made possible when the two superpowers reached a situation of strategic equilibrium and a relationship of balance of terror (Selim 2000: 137).

In addition, the Arabs did not view CBMs as a viable strategy for achieving peace in the Middle East. From an Arab perspective, the conditions that led to the success of CBMs in Europe in the 1970s did not exist in the Middle East. In Europe, when CBMs first emerged in 1975 in the context of the CSCE process, they were an outcome of a number of strategic developments that had occurred in East–West relations by the late 1960s and early 1970s. Perhaps the most important among these developments was the formalization of the territorial status quo in Europe and the normalization of inter-European relations. This process started in 1967 when Romania and Yugoslavia formally recognized West Germany, and eventually gained momentum by the early 1970s when West Germany, under Chancellor Willy Brandt's *Ostpolitik* doctrine, began to make rapid progress in normalizing relations with its eastern neighbours and the Soviet Union.[15] This manifested in the conclusion of a number of treaties that formalized the post-World War II territorial map of Europe.

In August 1970, West Germany and the Soviet Union signed the Treaty of Moscow, which legitimized the territorial and political consequences of World War II. The treaty included a non-aggression pact by which both sides renounced the use of force against each other as a means of settling any possible disagreements. It also included a declaration by West Germany that it had no territorial claims against any country and that it would regard today and in the future the frontiers of all countries in Europe as inviolable, including the frontier between West Germany and Poland and the one between West and East Germany (Hanrieder and Auton 1980: 67–68).

The Moscow Treaty opened the way to negotiations between West Germany and other countries of the Eastern bloc. In December 1970, West Germany concluded the Warsaw Treaty with Poland. The treaty formally renounced West Germany's territorial claims vis-à-vis Poland by recognizing the frontier between the two countries at the Oder–Neisse line, which had been in effect since the end of World War II. The two countries affirmed the inviolability of their existing frontiers, agreed to respect each other's territorial integrity without restriction, and declared that they had no territorial claims whatsoever against each other. In December 1970, West Germany and Czechoslovakia concluded the Treaty of Prague, which normalized relations and formalized the territorial status quo between the two countries. This was followed by the establishment of diplomatic relations between West Germany and Hungary and Bulgaria (Ortmayer 1975: 115).

In addition to the formalization of inter-state frontiers in Europe, the early 1970s also witnessed a settlement to the Berlin question through an agreement signed between the Soviet Union and the Western powers acknowledging the division of Berlin and recognizing the status of West Berlin as an area linked to, but officially separate from, West Germany. In September 1971, the four occupying powers of Berlin—the Soviet Union, the United States, France, and Britain—signed the Quadripartite Agreement, which stipulated that neither party would change the status quo in Berlin unilaterally, recognized each of the four powers' existing rights in their respective sectors in Berlin, and affirmed the special political relationship between West Berlin and West Germany.

---

15 The origin of these developments dates back to 1966 when a new governing coalition came to power in West Germany, bringing to the foreign ministry the Social Democratic leader, Willy Brandt, who was a strong advocate for the establishment of peaceful and normal relations with the Eastern bloc. Once in power, Brandt introduced his foreign policy doctrine of *Ostpolitik*, which meant pursuing a proactive policy towards the East. Under Brandt's doctrine of *Ostpolitik*, West Germany witnessed a foreign policy reorientation by seeking the normalization of its relations with its East European neighbours. The rationale was to adopt a more realistic foreign policy that would face up to the realities of the situation after World War II with its consequences on the ground. In 1969, West Germany's *Ostpolitik* received a strong boost when the October 1969 elections brought Brandt to the head of the German government as chancellor in a coalition government dominated by his Social Democratic Party. See Urwin 1997: 158–160.

Perhaps the most important achievement of Brandt's *Ostpolitik* was the signing of the Basic Treaty between the two Germanys in December 1972. According to the Basic Treaty, West Germany accepted the reality of East Germany as a sovereign state, thus dealing with it on the basis of full equality; renounced previous claims that only West Germany could legitimately speak for all Germans; and treated the frontier with East Germany as an inviolable political border rather than a demarcation line. The treaty also stipulated that ties between West Berlin and West Germany would be maintained and developed, taking into consideration that West Berlin was not a constituent part of West Germany and was not governed by it; and that Western powers would represent West Berlin abroad in all matters concerning security and status. Further, the treaty allowed for the development of commercial, cultural and personal relations as well as the exchange of permanent diplomatic missions between the two countries (Dean 1988: 83–105).

With the conclusion of these treaties, a new political environment was created in Europe. The political division of Europe, which had precipitated the Cold War between the two victors in the struggle against Nazi Germany, was finally acknowledged as a fait accompli. Not only was there mutual recognition of the consequences of World War II and the existing political realities on the ground, but there was also a mutual concern to preserve the territorial status quo between the two blocs and to work towards establishing normal relations across different fields. Taken together, these treaties provided Europe with what it had lacked since 1945: a post-war peace settlement based on the acceptance of the political and territorial status quo that had resulted from the continent's division.

The second development that enabled the introduction of CBMs was the American–Soviet détente which began with the launching of the SALT negotiations in Vienna and Helsinki in 1969. The American–Soviet arms control talks became possible after the reaching of the balance of terror in the early 1960s, when the two superpowers possessed the nuclear bomb as well as the delivery vehicles necessary for carrying the bomb to its targets such as Intercontinental Ballistic Missiles (ICBMs) and long-range bombers.[16] In September 1971, the American–Soviet détente gained momentum with the conclusion of the Agreement on Measures to Improve the Direct Communications Link in order to improve communications between the two countries in periods of crisis, and the Agreement on Measures to Reduce the Risk of Outbreak of Nuclear War to prevent the risk of a nuclear war arising from unauthorized or accidental use of nuclear weapons. One year later, détente was officially recognized with the historic visit of President Nixon to the Soviet Union in May 1972; he thus became the first serving American president to set foot inside the Kremlin. At this summit, the two

[16] The Soviet Union had produced the nuclear bomb in 1949, but did not possess the necessary delivery vehicle for carrying the bomb to the territories of the United States until 1957, when it produced its first ICBM. Prior to that date, there had existed a balance of power which rested on the capability of the Soviet Union to carry the nuclear bomb to western European allies of the US within NATO. But it was only by the early 1960s that the balance of terror between the two superpowers was fully reinforced.

superpowers concluded a joint declaration of principles by which détente and peaceful coexistence could be achieved and sustained. They agreed to "respect each other's sovereignty, equality and right to non-interference in internal affairs by the other, not to seek advantages, to avoid military confrontation, and to adhere to the renunciation of the use or threat of force" (Urwin 1997: 202).

During the summit, the two superpowers also concluded the Interim Offensive and SALT I agreements, which made significant achievements on the road to structural arms control between the two of them. One the one hand, the Interim Offensive Agreement limited the number of missile delivery systems with nuclear warheads for each party. A ceiling was imposed on the number of ICBMs that each side could deploy over a 5 year period, starting from October 1973. Another restriction was imposed on the construction of SLBMs over an equivalent period. On the other hand, the SALT I agreement limited the number, deployment and activities of anti-ballistic missile (ABM) defence systems for each side. The agreement stipulated that both sides would refrain from deploying ABM systems for the defence of other countries or for purposes of regional defence. The deployment of ABM systems was restricted to two sites for each, one to be the national capital and the other to be an ICBM missile base. The agreement also stipulated a maximum of 100 ABM launchers and interceptor missiles at each. With respect to the activities of ABM, the agreement banned mobile ABM systems and prohibited the construction of sea-based, air-based, space-based, or mobile land-based components (Farley 1988: 217–218).

It was in this context that the CSCE process was launched and CBMs were introduced in Europe by the mid-1970s. In Europe, the rise of CBMs was facilitated by the settlement of inter-European territorial disputes and the rise of East–West détente. Without such developments, the idea of the CSCE and CBMs "would have been unthinkable" (Maresca 1988: 107). Accordingly, the Arabs contend that CBMs in Europe did not trigger a change in the strategic environment in which they were applied, nor did they resolve inter-European disputes, as the process of territorial settlement had already been achieved prior to the introduction of CBMs. Rather, the CSCE process, alongside the CBMs adopted in Helsinki and other subsequent frameworks, was meant to institutionalize and legitimize the newly emerging geostrategic reality in East–West relations by the mid 1970s. It was meant to provide a multilateral ratification of the treaties reached between West Germany and its East European neighbours on the recognition of political frontiers, and to institutionalize such treaties under a regional pan-European forum in a way that gave the process a degree of irreversibility. In this process, CBMs were developed as a mechanism oriented to the status quo whose purpose was to stabilize military relations between NATO and Warsaw Pact members. This was achieved by providing each side with tangible and verifiable assurances regarding the purpose and character of military activities of the other, thus preventing surprise attacks or war by misperception. In other words, CBMs were developed as a post-conflict resolution mechanism, which means that their objective was conflict prevention, and not conflict resolution.

In the Middle East, however, the region has lacked most of the enabling conditions necessary for the application of genuine CBMs. Unlike in Europe, accepted legal borders still remain a key source of antagonism between the Arabs and Israel today. Despite the launching of the Middle East peace process in the 1990s, the major territorial issues between Israel and the Arab countries are still unresolved. With the exception of Egypt and Jordan, Israel is still in a formal state of war with her Arab neighbours. In Palestine, Israel still occupies most of the Palestinian territories it captured in 1967, and has formally annexed East Jerusalem. The Oslo peace process did not lead to the settlement of the Israeli–Palestinian conflict. Although the 1993 DOP referred to a settlement on the basis of Security Council Resolutions 242 and 338, these resolutions were scrapped during the negotiations. Whereas the Palestinians understood the peace process as a means to implement the UN resolutions, Israel expected a settlement that reflected the power imbalance, rather than the mutually agreed legal international framework. In fact, the negotiations proceeded on the basis of the balance of power between the Palestinians and the Israelis, which was definitely in Israel's favour. These two different conceptions were a determinant factor in the failure of the Oslo peace process, which came to an end with the outbreak of the second Palestinian Intifada in September 2000. In Lebanon, the southern part of the country was occupied by Israel for almost 18 years (1982–2000). During this period, Israel confronted heavy pressure from Hezbollah in southern Lebanon and was forced to station more forces in that region as its Lebanese allies were faltering. Although Israel took an audacious decision in 2000 to withdraw from southern Lebanon, the withdrawal was not part of a settlement agreement with the Lebanese government. Rather, it was a unilateral withdrawal justified by security considerations. Further, Israel left one enclave, Shaba's Farms, under Israeli control, and this led Hezbollah to claim that the withdrawal was not complete. In Syria, the series of Syrian–Israeli peace talks conducted within the framework of the Madrid process did not result in a settlement of the question of the Golan Heights. This was due to Israel's reluctance to carry out a complete withdrawal to the borders of 4 June 1967. Taken together, the Israeli occupation of Arab territories, coupled with the lack of progress in the Arab–Israeli peace process, has restricted attempts to apply CBMs between the two sides. As a mechanism oriented towards the status quo, CBMs could not be applied to a situation where one of the concerned parties is not satisfied with the status quo.

Similarly, the emphasis of the Mediterranean Charter on conflict prevention rather than conflict resolution reflects an approach oriented towards the status quo. In fact, the projected charter was called the Charter for Stability. The notion of stability can only be accepted if there is an agreement among the actors on the basic parameters of the situation, as was the case in Europe in the mid-1970s. In the eastern Mediterranean region, there is no agreement on these parameters. Under these conditions, an emphasis on the notion of stability would serve the interests of some actors to the detriment of the others. For example, an emphasis on stability and conflict prevention in the eastern Mediterranean would provide Israel with ample time to absorb the Arab occupied territories during which it would not be disturbed by Arab resistance to occupation. Under these conditions, it is difficult to speak meaningfully of a security partnership.

Accordingly, one can understand the limitations of global CBMs proposals in the Middle East in the post-Cold War era. The nature of CBMs as status-quo-oriented and post-conflict resolution measures has made the Arabs reluctant to get involved in CBMs arrangements with Israel at a time when parts of their territories are still occupied by Israel. From an Arab perspective, the main problem with the Western CBMs initiatives was that they virtually bypassed the core issue in the Arab–Israeli conflict; the occupation of Arab territories. Under a condition of territorial occupation, the Western CBMs proposals, the Arabs argued, would mean in practice the perpetuation of Israeli hegemony, both politically and militarily, in the region. The proposals, had they been adopted and implemented, would have led to the legitimization of the territorial status quo manifested in the Israeli occupation of Arab territories.

In the third Euro-Med ministerial meeting held in Stuttgart in 1999, the then Egyptian foreign minister, Amr Moussa, outlined the Arab position on the security dimension of the Barcelona process, arguing that "The Barcelona process is not the Middle East peace process. It is not a negotiating framework, but a larger framework dealing with cultural, economic, and security issues. Among these issues are the threats to Middle East peace process… These threats influence the Barcelona process. Consequently, the Barcelona process must pay attention to the major problems in the Mediterranean such as the Middle East and Kosova". In fact, Moussa made it clear that Arab countries are not likely to proceed with the enforcement of the projected Mediterranean Charter unless the Middle East peace process is resolved.[17] In a meeting held by the eight Arab partners in the EMP in Syria in August 1999 to coordinate policies towards the projected charter, the Arabs agreed that "the charter will be implemented only after the settlement of the Arab–Israeli conflict and the achievement of a just and comprehensive peace in the Middle East".[18] In 2000, El-Shazly (2000: 27), the then Egyptian Ambassador to Turkey, summarized the Arab position as follows:

> It was widely believed among Arabs that no process entitled peace and stability could be embarked upon between partners while some of them were legally in a state of war. Arabs also believed that military confidence and security building measures under these conditions would practically amount to bestowing blessing and tolerance on the foreign occupation of Arab territories.

Thus it was quite normal that the ACRS working group activities stagnated with the collapse of the Middle East peace process in 1996. In this year, the ACRS did not hold any meetings, and in 1997 the Arab League decided to freeze its participation not only in the ACRS, but also in the Madrid Multilateral Track as a whole. Similarly, the NATO—Mediterranean Dialogue achieved limited success, and few of the proposed CBMs were actually implemented. This limited success

---

17 Interview with Amr Moussa, *Al-Ahram* (Cairo), 16 April 1999.

18 Statements by Ambassador Fathy El-Shazly, Egypt's Assistant Minister for Foreign Affairs, summing up the conclusions of Arab high officials participating in the Coordinating Meeting held in Damascus. *Al-Ahram* (Cairo), 19 August 1999.

was the result of several factors; chief among them were (i) the misgivings of the Arab Mediterranean countries about NATO as a Cold War alliance, (ii) Arab perception of the Dialogue as a means of entering into cooperative security schemes with Israel and monitoring their military forces and activities while Israel still occupied parts of their territories, and enjoyed military superiority over the Arab states in both conventional and unconventional weapons; and (iii) the deterioration of the Middle East peace process (Larrabee et al. 1998: 80). The shift of NATO after 9/11 in the direction of playing a military role in support of the American military occupation of Afghanistan also had a negative impact on the hopes that such CBMs would actually be implemented.

5.2 Second Stage: Declining Arab Interest in Arms Control

In the wake of 9/11, Arab conceptualizations of arms control underwent fundamental changes which were reinforced after the United States emerged in the Arab world, for the first time, as the military occupier of an Arab country in 2003. The invasion of Iraq brought the United States from a traditional external influencer of Middle East politics to a key regional player through its military and political presence in Iraq. This provided the United States with the opportunity to develop a more detailed and invasive policy for the Middle East region than ever before, and this in turn had immediate consequences for the question of arms control and counter-proliferation in the region.

One of the direct consequences of the Bush pre-emptive doctrine and the invasion of Iraq was that they brought most Arab countries under 'existential' pressures to cooperate with the American projects in the region or face assured collapse. In the lead-up to the invasion, the United States put intensive pressure on Arab regimes to join the Anglo-American campaign in Iraq. Although Arab regimes were initially reluctant to endorse the American plans, with the exception of Syria and Arab Maghreb countries they finally succumbed and joined the Anglo-American campaign with varying degrees of publicity. In fact, former Egyptian president Hosni Mubarak had argued before the invasion that Iraq had WMD and it should abide by UN resolutions. Egypt and the GCC states also provided the invading Anglo-American forces with critically important military facilities, including access to military bases, water passages and airspace. Following the invasion, the United States set the tone for a new Middle East. In April 2003, the then US Secretary of State, Colin Powell, told reporters in Washington that a new strategic situation is currently emerging in the Middle East, and that all countries of the region must reconsider their previous policies on the basis of the new situation resulting from the war in Iraq.[19] John Bolton also indicated frankly that other

[19] *Al-Ahram* (Cairo), 15 April 2003.

countries in the region might become possible targets for US disarmament efforts, stating that:

> We [the United States] are hoping that the elimination of the dictatorial regime of Saddam Hussein and the elimination of all of Iraq's weapons of mass destruction would be important lessons to other countries in the region, particularly Syria, Libya, and Iran, that the cost of their pursuit of weapons of mass destruction is potentially quite high. We want a peaceful resolution to all of these issues, but the determination of the United States, especially after September 11, to keep these incredibly dangerous weapons out of the hands of very dangerous people should not be underestimated.[20]

Arab vulnerability to American pressure was further reinforced by the fact that the majority, if not all, pro-American Arab regimes derived their legitimacy mainly from American recognition of their role as strategic local clients responsible for implementing the American agenda in the Middle East. This external legitimation went in parallel with a crisis of internal legitimacy that characterized most of these regimes, and which resulted from systematic state oppression against local civil society and opposition groups. The Arab regimes legitimized their use of terror and the persistent violations of basic human rights on the grounds that they were simply maintaining stability and order. In addition, internal legitimacy was also eroded as a result of the persistence of various forms of economic and social injustice in most of these countries. Almost a third of Arab youth were unemployed, most of them university graduates. The Arab regimes failed to achieve genuine development and mostly depended on external rent and the selling of public assets at extremely low prices to domestic and external clients of the regimes. Most national wealth was either unfairly sold or simply smuggled to foreign banks (Shehata and Wahid 2011: 10–17). The limited development which was achieved was directed mainly to the rich elite in alliance with the ruling ones. In fact, the gap between rich and poor and the control of national wealth by a limited minority of businessmen was never more apparent in any other stage of recent Arab history. In this context, the state–society tension that marked the contemporary Arab state system made American support largely indispensible for most Arab regimes to secure their survival and maintain their monopoly of political power in such troubled domestic settings.

Under such existential threats, Arab regimes were reluctant to articulate anti-American views on the new counter-proliferation agenda. The survival instinct led most of them to express their views cautiously in order not to offend the Bush administration and as a result lose American support or become a target of US wrath. In fact, out of survival instinct, some Arab regimes changed their foreign policies in the direction of acknowledging the American approach to regional politics, including the American WMD agenda. This change was evident at several levels.

At a general level, Arabs dropped their long-standing emphasis on the centrality of the Arab–Israeli conflict, traditionally viewed as the most acute threat to Arab

---

20 Radio Sawa Interview with Under Secretary John Bolton, 5 April 2003; available at: <http://iipdigital.usembassy.gov/st/english/texttrans/2003/04/20030416160433samohtj0.1897852.html#axzz2ChZI55TZ>. .

regional security. Following the invasion of Iraq, the United States strove to turn the Arab regional understanding against Iraq into an Arab–Israeli tacit understanding against Iran, with Egypt, the GCC states, and Jordan being the most important Arab actors in these understandings, and Israel being an undeclared member. The Arab states which joined the American-led regional understanding were even willing to enter into normalization initiatives with Israel, although their conflict with Israel had not been resolved. This was reflected in their participation in the Annapolis Conference held in 2007 and their endorsement of the participation of the Arab League in the Union for the Mediterranean, in association with Israel. This was the first time the League took part in a regional arrangement in which Israel participated. In May 2004, Egypt concluded its largest economic deal ever with Israel: a contract worth US$2.5 billion for Egypt to supply Israel with natural gas at price levels below the global price levels. One year later, Egypt further enhanced its cooperation with Israel with the conclusion of the *Qualified Industrial Zone* (QIZ) agreement, by which Egyptian goods with Israeli origins would gain free access to US markets.

In the same vein, the Arab states dropped their traditional emphasis on the crucial role of Palestinian armed resistance to Israeli occupation. This explains the Arab boycott of Hamas, a radical Islamist organization advocating armed resistance against Israeli occupation, when it was elected to form the Palestinian government in 2006. With the outbreak of the rift between Mahmud Abbas, the head of the Palestinian Authority, and the Hamas-led Government of Ismail Haniya, and the latter's control of the Gaza Strip in 2007, Egypt and Israel imposed an economic blockade on the Gaza Strip hoping to bring about the downfall of that government. Both of them blocked all crossing-points from Egypt and Israel into the Gaza Strip; this included blocking humanitarian supplies. The Israeli military offensives against Hezbollah and Hamas in 2006 and 2008 respectively helped clarify these perceptions. In 2006, some Arab governments, mainly Egypt, Jordan, and Saudi Arabia, accused Hezbollah of provoking the Israeli military onslaught on Lebanon. In a joint statement during the Israeli aggression against Lebanon, King Abdullah of Jordan and Egyptian President Mubarak put the blame on Hezbollah for the outbreak of the war by referring to "uncalculated adventures that do not serve the interests of the region". The joint statement also made indirect reference to the need to implement UN Security Council Resolution 1559 which called, among other things, for the disarmament of Hezbollah. The Saudi government also condemned Hezbollah's actions as 'irresponsible adventurism'.[21] In addition, in 2008 the three Arab countries expressed an 'understanding' of the motives of the Israeli onslaught on the Gaza Strip by blaming Hamas, and used the onslaught to introduce measures which would topple Hamas from its rule in Gaza.

In the area of arms control, Arabs dropped their emphasis on the reciprocity of commitments in the area of regional arms control. Some Arab regimes even became more willing to adhere to the US counter-proliferation doctrine in order to

[21] *Al-Ahram Weekly* (Cairo), Issue No. 804, 20–26 July 2006.

ensure survival. An obvious example here was the decision of the then Libyan leader, Qaddafi, to abandon his country's WMD programmes and cooperate fully with American and British weapons inspectors in order to secure the survival of his regime in a post-Saddam Hussein, American-dominated regional order. In December 2003, Libya announced it would dismantle its WMD programme and open the country to immediate and comprehensive verification inspections. Libya pledged to "eliminate all elements of its chemical and nuclear weapons programs; declare all nuclear activities to the IAEA; eliminate ballistic missiles beyond 300 km [kilometres] range, with a payload of 500 kg [kilogrammes]; accept international inspections to ensure [its] complete adherence to the Nuclear Nonproliferation Treaty, and sign the Additional Protocol; eliminate all chemical weapons stocks and munitions, and accede to the Chemical Weapons Convention; [and] allow immediate inspections and monitoring to verify all of these actions".[22] In March 2004, Libya transferred all technical data, material, and weapons production equipment for its nuclear programmes to the United States and allowed full IAEA inspection. It also ratified the NPT and the *Comprehensive Test Ban Treaty* (CTBT), and acceded to the CWC and the BWC although it had not developed or produced biological weapons. In this process, as reported by *The New York Times*, Qaddafi "allowed CIA and other American nuclear experts into the country to assess Libya's equipment and bomb designs and to arrange for their transfer out of the country".[23] This was done despite Libya's traditional emphasis on the Israeli nuclear arsenal as a security threat. In fact, while the Libyans called upon Iran to pursue the Libyan disarmament model, they did not call upon Israel to do likewise. As expressed by one Egyptian scholar:

> Qaddafi decided to change his skin, in the hope of winning the approval of the masters of the new era, even if this means he has to go full circle and lead the counter-revolution against himself, at his own pace of course, in order to perpetuate his regime in its new guise. Only in this context it is possible to understand his decision to open his country up to inspection and show his willingness to comply with everything Washington and London ask (Nafaa 2004).

Arab reactions to the Israeli strike on Syria in 2007 also illustrated these perceptions. The strike was met with a near total silence and lack of criticism in the Arab world, including Syria itself. The Syrian initial reaction did not go beyond a complaint that Israeli aircraft had violated its airspace and dropped some explosive charges in a remote, desolate area. Instead of protesting against the Israeli aggression, the Syrian regime tried in fact to downplay the issue in an attempt to avoid any further escalation with Israel or the United States, and to justify its reluctance to retaliate. Although the Syrian regime complained immediately about Israel's

[22] The White House, Fact Sheet: *The President's National Security Strategy to Combat WMD, Libya's Announcement*, 19 December 2003; available at: <http://2001-2009.state.gov/p/nea/rls/27462.htm>.

[23] "C.I.A. Agents in Libya Aid Airstrikes and Meet Rebels". *The New York Times*, 30 March 2011; available at: <http://www.nytimes.com/2011/03/31/world/africa/31intel.html>.

unauthorized intrusion into its air space, it initially claimed that the warplanes had rapidly retreated back to Israel after they encountered Syrian air defences (Butcher 2007). Subsequently, Syrian officials stated that the Israeli warplanes had dropped their munitions on Syrian territory, but had either failed to hit a concrete target or had simply destroyed an empty warehouse (Cooper and Mazzetti 2007). A few days after the incident, Syrian foreign minister Walid Moallem showed European diplomats alleged photographs of the target site in order to support his contention that the Israelis had struck nothing (Boudreaux and Daragahi 2007). Syrian Vice-President Farouq Al-Shara went further and called for closing any discussion on this issue, warning that "Those who continue to talk about this raid and to invent inaccurate details are aiming to justify a future aggression [against Syria]".[24] In his first comments on the topic, Syrian President Bashar Al-Assad told *BBC News* on 1 October that a Syrian 'unused military building' was attacked by Israel, but no further details were provided. Although Al-Assad warned that "Syria reserves the right to retaliate to the attack", his remarks left it ambiguous whether retaliation might involve a military counter-strike as opposed to diplomatic countermeasures. "Retaliate does not mean missile for missile and bomb for bomb. We have our means to retaliate, maybe politically, maybe in other ways. But we have the right to retaliate."[25] In subsequent statements, Syria denied American allegations that the Israeli strike had targeted a covert nuclear facility in Syria. In June 2008, the Syrian government agreed to allow an IAEA inspection team into the site to validate Israeli and American allegations.

At the regional level, the silence was even more acute, as all Arab governments responded with 'no comment'. As explained by one Arab analyst, "There was no condemnation and no solidarity from the Arab front, unlike other similar cases of Israeli aggression in 2001 and 2003… Major Arab satellite channels and newspapers (all funded by or close to the Saudis) even adopted a stance one step short of holding the Syrians responsible for the Israeli act" (Moubayed 2007). In fact, the only countries to condemn the Israeli attack were from outside the Arab world, including Russia, Turkey, and North Korea, who all came out with a severe condemnation of Israeli aggression. The Arab silence reflected a deep sense of helplessness on the part of Arab ruling elites in the face of the US counter-proliferation agenda.

In the same vein, Arabs also dropped their long-standing emphasis on the linkage between endorsing the Additional Protocol of the NPT and Israel's endorsement of the NPT. In this context, Iraq, Kuwait, and Libya had already ratified the Additional Protocol, and were joined later by the UAE, which signed the Protocol in April 2009. This was despite the fact that neither Kuwait nor the UAE possessed any nuclear programmes or research-related activities at the time of ratifying the Protocol. In the same vein, Libya ratified the CTBT and CWC, and agreed to more

---

[24] "Syria Says Israeli Air Raid Aimed at Justifying Attack". *Agence France Press*, 29 September 2007; available at: <http://afp.google.com/article/ALeqM5hcd8yNVwB0Z6ZOjoi1YKbiWIremw>.

[25] "Assad Sets Conference Conditions". *BBC*, 1 October 2007; available at: <http://news.bbc.co.uk/2/hi/middle_east/7021986.stm>.

stringent IAEA inspections. In addition, Egypt, Jordan, and the GCC declared their intention to pursue peaceful nuclear programmes that would eventually require them to join the list of signatories for the Additional Protocol. As explained by one Egyptian scholar, such declarations would not go beyond official rhetoric and public statements as they did not reflect a genuine Arab desire to acquire the nuclear technology. Rather, they were mechanisms designed by Arab governments to rationalize, for their public opinion, their intention to endorse the Additional Protocol, as signing the Protocol was a prerequisite for proceeding with their nuclear programmes (Selim 2008).[26]

Parallel with their de-emphasis on Israel's nuclear programme, Arab regimes, with the exception of Syria, perceived Iran's nuclear programme as the main threat to their national security. Arab perceptions of the Iranian nuclear programme were articulated in the context of two main regional developments. The first was the Western (particularly American) drive to persuade or coerce Iran to halt its programme, in line with Western attempts to isolate Iran through the mobilization of Arabs against its nuclear programme. Before the launching of the Western anti-Iranian drive, virtually no Arab state had voiced a concern over the Iranian nuclear programme. In fact, Arab states had recognized Iran's right to develop a civilian nuclear programme, and they had no reason to doubt its peaceful intentions. However, the Western strategy aimed to create a coalition between Israel and pro-Western Arab states against Iran and its allies in the region. This strategy was implicitly acknowledged in the 2007 Riyadh Arab Summit, which laid the foundation for the political division of the Middle East into (i) a 'radical' camp, including the indigenous forces of resistance, mainly Iran, Syria, Hezbollah, and Hamas, and (ii) a 'moderate' camp, including American allies, mainly Egypt, Jordan, the GCC states, the Palestinian Authority, and Israel.

The second development had to do with the rise of Iranian political influence in Iraq and the Middle East. Following the US invasion of Iraq in 2003, the United States established a new political order in that country based on a sectarian quota system through which the Shi'a political factions, many of which were based and trained in Iran, controlled the Iraqi political scene, thus ending the monopoly of Sunni Arabs over Iraq's governing structures. This trend was viewed as a disturbing development by many Sunni Arab states, a fact explained in light of the historical Sunni–Shi'a rift over the nature of the Islamic political-religious order following the death of Prophet Mohammad.[27] The Sunni–Shi'a conflict had been indelibly imprinted on the history of the region, especially after Persia (currently Iran) turned to Shi'ism in 1500. For centuries, the conflict took the form of wars and invasions between the Ottoman Empire and the various Persian dynasties. However, this

---

[26] For a similar remark, see interview with Hisham Fu'ad, the former chairman of Egypt's Atomic Energy Agency. *Al-Masry Al-Youm* (Cairo), 10 April 2009.

[27] Whereas the Sunnis believe that the leader of the Islamic state should be freely chosen by the believers and be held accountable to the Muslim Umma (nation), the Shiites believe that such leadership (Imam) is the prerogative of the Prophet's cousin Ali Ibn Abi-Talib and his sons, and in fact the caliphs who succeeded Prophet Mohammad had usurped power.

sectarian rift was resurrected after the Iranian revolution of 1979, and became a major feature of Middle East politics after the Anglo-American invasion of Iraq. For the majority of the Sunni Arab states, the rise of Shi'a influence in Iraq was triggered by persistent Iranian intervention in the post-Saddam Iraqi political scene. Although dominated by Sunni majorities, many Arab states are home to significant Shi'a minorities, which are perceived as having strong connections to Iran as the most prominent Shi'a country. In this respect, the Arab Sunni states felt that the rise of Iranian influence in the Middle East would lead to the empowerment of the Shi'a minorities in the Arab world, and this in turn would threaten the security of Arab regimes and undermine their control over their people. This led the majority of Arab regimes to view an ascendant Iran as representing a major source of threat to their domestic security, and gave these regimes a powerful incentive to try to limit Iran's political influence in the region, including its nuclear programme. In Jordan, King Abdullah II warned of an emerging 'Shi'a Crescent' extending from Iran to Lebanon, where Hezbollah, a Shiite organization, was on the rise. The Saudis also subscribed to these perceptions, although they were more restrained in articulating them. They nevertheless intervened in Lebanese politics to support Sunni forces against Hezbollah. In Egypt, Mubarak laid the charge that the loyalty of all Shiites, Arabs and non-Arabs, was to Iran, a statement that outraged Arab Shiites.[28]

Arab perceptions of the Iranian nuclear programme are best illustrated in the announcement of two non-proliferation initiatives in 2005 and 2007 respectively. In the GCC summit meeting held in Abu Dhabi in December 2005, the GCC Secretary General Abdel-Rahman Al-Attiyah announced an initiative to declare the Gulf a WMDFZ. The initiative was based on the principle of progress from a sub-regional to a regional approach. It called for the establishment of a WMDFZ in the Gulf region as a first stage toward the realization of a comprehensive WMDFZ in the Middle East at a later stage. The initiative defined the Gulf region as the area consisting of the nine states of the geopolitical Gulf region, namely Bahrain, Oman, Kuwait, Saudi Arabia, Qatar, UAE, Iraq, Iran, and Yemen.[29] Accordingly, the initiative would oblige these states to refrain from the development of WMD and to remove any stockpiles of these weapons, while other regional states would be under no obligation to disarm. Given that Iran has been the only state in the Gulf region with a nuclear programme, one can understand the real objective of the initiative, which is to inhibit Iran from the possible development of nuclear weapons in the future.

In the same meeting, the GCC states strongly criticized Amr Moussa, the Arab League's Secretary General, for his critical remarks on the initiative. In a letter to the GCC summit, Moussa had articulated the following view on the initiative:

> The GCC project, if it achieves any success, represents a process to undermine collective Arab efforts that have been exerted to stipulate a draft covenant to make the entire Middle

[28] "Egypt's head questions Shiites' loyalty". *The Michigan Daily*, 10 April 2006; available at http://www.michigandaily.com/content/egypts-head-questions-shiites-loyalty.

[29] "The Declaration of the Gulf WMDFZ Initiative by the GCC Secretary General". *Al-Sharq Al-Awsat* (Riyadh), 19 December 2005.

East region free of all WMD. It also confuses Arab efforts on the international scene to implement this initiative.

The security of the region, especially with respect to WMD, is integral and indivisible, and dealing with this question in the Middle East region must take place in the context of a comprehensive regional vision of security and stability, not via selective dealing, whether on the basis of a sub-regionalism that lacks strategic depth, or in a selective manner that corroborates the possession of nuclear weapons by one state—Israel.

The attempt to launch the idea of founding a region free of WMD in the Gulf region gives the impression that Gulf security is not linked to achieving security in the Middle East and consequently Gulf is not affected by the acquisition of Israel of nuclear weapons.[30]

The GCC states criticized Moussa publicly on the grounds that he had ignored the security concerns of the Gulf region. In his response to the letter from Moussa, Abdullah Al-Neamey, the foreign minister of the UAE, stated:

We appreciate the fears and worries of Mr. Amr Moussa, as an Egyptian, as regards the Israeli nuclear programme…We in the Gulf also harbour fears and worries with respect to the Iranian nuclear programme. When speaking of Arab national security, Amr Moussa was supposed to speak of the fears in the Gulf since the GCC States are near to the Iranian nuclear reactor in Bushehr. We do not have protection or prevention measures in case there is a leakage from the reactor especially since Iran is not a signatory state to the Early Warning Convention…The worries of GCC States about the Iranian nuclear reactor are legitimate. Amr Moussa does not share these worries with us; he is free to do so…We hope that when the Secretary General speaks to the Arab League of the dangers threatening the region, he takes into consideration the existence of six Arab Gulf States which are part and parcel of the Arab security system.[31]

In October 2007, the Saudi foreign minister Saud Al-Faisal put forward another GCC initiative which invited all states of the region to participate in the establishment of a Uranium Enrichment International Consortium to be based in a neutral country outside the region. The project sought to establish a joint enrichment and processing centre to supply nuclear fuel to civilian reactors in the Middle East region. Participants in the Consortium would thus secure the supply of nuclear fuel for their nuclear power plants, but they would forego developing or completing uranium enrichment programmes on their own territories. The GCC proposal aimed to centralize enrichment activities to prevent regional actors from militarizing their civilian programmes and to ward off the possibility of a nuclear arms race in the region.[32] It is obvious that the GCC proposal was mainly targeting Iran's attempts to develop a national uranium enrichment technology for its nuclear programme. Meanwhile, the proposal would have no effect on Israel, which had already mastered the enrichment technology a long time ago, and was already in possession of hundreds of nuclear warheads. In March 2009, Al-Faisal articulated the GCC position more bluntly at an Arab League Council meeting. He called on his Arab counterparts

30 "The Letter of the Secretary-General of the Arab League to the Secretary-General of the GCC Objecting to the Gulf WMD Free Zone Project". *Al-Hayat* (London), 2 January 2006.

31 *Asharq Al Awsat* (Riyadh), 20 December 2005.

32 "Gulf states 'offer Iran uranium'." *BBC News*, 1 November 2007; available at: http://news.bbc.co.uk/2/hi/middle_east/7073699.stm .

to forge a common vision to deal with what he called the 'Iranian challenge', arguing that resolving problems among Arabs would depend on the creation of a joint position regarding Iran's stance on Gulf security and its nuclear programme.[33]

In July 2010 and in unusually candid remarks, the UAE Ambassador to Washington Yousef Al-Otaibi publicly endorsed the 'military option' for countering Iran's controversial nuclear programme, arguing that the benefits of bombing Iran's nuclear facilities might outweigh the costs. In a public interview session with *The Atlantic* magazine, Al-Otaibi called Iran "the only country to pose a threat to the UAE in the region", and emphasized that "it is in [his] country's best interest that Iran does not gain nuclear technology". Al-Otaibi voiced scepticism about the value of deterrence and containment, noting that "Iran thus far has not been deterred from other aggressive activity like supporting terrorist groups such as Hezbollah". He even urged the United States to stop the Iranian nuclear programme by force, arguing that the UAE was "at risk of an Iranian nuclear program far more than [the US is] at risk. At 7,000 miles away, and with two oceans bordering you, an Iranian nuclear threat does not threaten the continental United States. It may threaten US assets in the region, it will threaten the peace process, it will threaten balance of power, it will threaten everything else, but it will not threaten you [the United States]."[34]

Egypt subscribed to a similar, though more cautious, view on the question of Israeli and Iranian nuclear programmes. In his speech before the April 2010 World Nuclear Summit in Washington, Egypt's foreign minister Ahmad Abul-Gheit did not refer to Israel's nuclear arsenal as part of the challenge facing the NPT regime nor to efforts to create a Middle East WMDFZ. Egypt's top diplomat not mentioning Israel in an international summit devoted to discussing measures for nuclear non-proliferation was considered a major transformation in Egyptian foreign policy, given that Egypt had frequently taken advantage of similar international platforms to mobilize against Israel and single out its nuclear programme as a major obstacle to the creation of a WMDFZ in the Middle East.[35] In December 2010, Abul-Gheit was blunter in articulating Egyptian perception of the Israeli and Iranian nuclear programmes. Speaking to the ruling National Democratic Party (NDP) annual conference, he purposely excluded Israel's nuclear weapons from the list of major challenges facing Egypt's diplomacy and national security. The other challenges Abul-Gheit mentioned were Sudan's stability, the Palestinian issue, Iran's nuclear programme, the Iraqi and Lebanese political landscapes, and engagement with global powers.[36]

---

[33] *Al-Ahram Weekly* (Cairo), Issue No. 938, 12–18 March 2009.

[34] "UAE Ambassador Endorses Military Force Against Iran Nuke Program". *CBSNEWS*, 7 July 2010; available at: <http://www.cbsnews.com/8301-503544_162-20009845-503544.html?tag=contentMain;contentBody>.

[35] "Egypt foreign minister ignores Israeli nukes at Washington summit". *Al-Masry-Alyoum* (Cairo), 12 April 2010.

[36] "Foreign Minister Abul Gheit delineates diplomatic challenges". *Al-Masry-Alyoum* (Cairo), 26 December 2010.

The release of thousands of confidential US diplomatic cables by Wikileaks corroborated Arab perceptions of a nuclear Iran. According to one leaked cable from the US embassy in Riyadh, in December 2005 Saudi Arabia's King Abdullah bin Abdel-Aziz expressed his anger that the Bush administration had ignored his advice not to go to war against Iran, arguing that "whereas in the past the U.S., Saudi Arabia and Saddam Hussein had agreed on the need to contain Iran, U.S. policy had now given Iraq to Iran as a gift on a golden platter". According to another leaked US cable, the Saudi King urged US General David Petraeus and Ryan Crocker, US ambassador to Iraq, in 2008 to "cut off the head of the snake" and take military action against Iran's nuclear programme. "May God prevent us from falling victim to their evil…We have had correct relations over the years, but the bottom line is that they cannot be trusted," Abdullah said, adding that "working with the US to roll back Iranian influence in Iraq is a strategic priority for the Saudi government".[37] Similarly, cables from the US embassy in Cairo revealed that Egyptian President Mubarak told US Senator John Kerry in June 2008 that Iran was a known state sponsor of terrorism and that he backed the idea of sanctions. Mubarak urged the United States to escalate its pressure on Iran, describing the Iranians as "big, fat liars" who "justify their lies because they believe it is for a higher purpose", and stressed that "this opinion is shared by other leaders in the region". In addition, cables from the US embassy in Bahrain revealed that King Hamad bin Isa Al Khalifa "argued forcefully for taking action to terminate Iran's nuclear program, by whatever means necessary". Hamad added, "That program must be stopped…The danger of letting it go on is greater than the danger of stopping it." The cables detail that "Bahrain had agreed to NATO's request to use Isa Airbase for AWACS missions, although the detail on numbers and timing have yet to be discussed".[38] Another leaked document also revealed that the UAE's defence chief, Crown Prince Bin Zayed, warned the United States that "a nuclear-armed Iran would destabilize the Gulf region and possibly allow terrorist access to WMD". According to the cable, Bin Zayed suggested in a meeting with the commander of the US Central Command, General John Abizaid, in 2006 that the United States must take action against Iran "this year or next", and wondered "whether it would be possible to 'take out' all locations of concern in Iran via air power". Bin Zayed even compared Iranian President Ahmadinejad to Hitler in urging against appeasing Iran.[39]

In sum, the leaked cables show the extent to which Arab regimes were alarmed by Iran's nuclear programme and its rising influence in the region. In none of these

---

37 "A Selection from the Cache of Diplomatic Dispatches". 19 June 2011; available at: <http://www.nytimes.com/interactive/2010/11/28/world/20101128-cables-viewer.html#report/iran-08RIYADH649>.

38 "A Selection from the Cache of Diplomatic Dispatches". 19 June 2011; available at: <http://www.nytimes.com/interactive/2010/11/28/world/20101128-cables-viewer.html#report/iran-09MANAMA642>.

39 "A Selection from the Cache of Diplomatic Dispatches". 19 June 2011; available at: <http://www.nytimes.com/interactive/2010/11/28/world/20101128-cables-viewer.html#report/iran-05ABUDHABI2178>.

cables, however, was nuclear Israel mentioned as a potential threat to Arab security. As reported by *The New York Times*, "The cables reveal how Iran's ascent has unified Israel and many longtime Arab adversaries—notably the Saudis—in a common cause. Publicly, these Arab states held their tongues, for fear of a domestic uproar and the retributions of a powerful neighbor. Privately, they clamored for strong action—by someone else."[40] Similarly, Hasanien Haykal, a prominent Arab intellectual, contended that the Sunni–Shi'a conflict has become the major feature of Middle East politics after the invasion of Iraq. According to his analysis, the nature of conflict in the Middle East has transformed from the traditional, territorial-based Arab–Israeli conflict to a newly emerging sectarian-based conflict between the Sunni Arab states on the one hand, and Iran, backed by Syria, Hamas, and Hezbollah, on the other.[41]

Nevertheless, these perceptions were not widely shared at the level of Arab civil society. The majority of Arab civil society groups rejected the US arms control agenda in the Middle East and viewed external pressure for counter-proliferation in the Arab world as an attempt to serve the interests of great powers and of Israel, rather than as real attempts to create a WMDFZ in the region. The rejectionists were a conglomeration of Arab nationalists, Islamists, leftists and some segments of the liberals, thus representing the mainstream view in the Arab world. All of them had in common either the rejection of globalization, which they perceived as equivalent to 'Americanization', or the advocacy of a cautious and conditional integration with its processes. In addition, they viewed hard security issues, such as the Iraqi and Palestinian questions and the Israeli nuclear programme, as the most crucial security issues facing the Arab world. They were also critical of Western military interventions and the American WMD agenda in the region, which aimed, according to their view, at reinforcing the strategic imbalance in favour of Israel. This was clearly manifested in the West's policy towards Iraq, which had been subjected to the most severe system of international sanctions in the twentieth century, leading to the misery of millions of Iraqis and culminating in the Anglo-American invasion of Iraq in 2003.

In fact, with the invasion of Iraq, the rejection by Arab civil society of American policy in the region was further magnified. We have argued elsewhere that the invasion of Iraq virtually pushed the Arab world back into the colonial era, where military power became the main tool of Western powers in achieving their interests in the region. Under such conditions, the Arab world became militarized as never before, as wars and invasions dominated the region in Iraq, Palestine, and neighbouring Afghanistan (Selim 2011: 325). In this context, the invasion of Iraq reaffirmed mainstream Arab perceptions of Western attempts to dominate the Arab world, which ultimately led to the rejection of all forms of Western intervention in the region, including the American arms control and counter-proliferation agenda.

---

40 "Around the World, Distress Over Iran". *The New York Times*, 28 November 2010; available at: <http://www.nytimes.com/2010/11/29/world/middleeast/29iran.html?_r=2&pagewanted=all>.

41 Interview with Mohammad Hasanien Haykal, *Al-Ahram* (Cairo), 21 May 2012.

In the same vein, Arab civil society has maintained positive views about Iran and its nuclear programme. According to the majority of Arab civil society, Iran is entitled to uranium enrichments under the NPT and should be allowed to use such a privilege. They also argued that the Iranian nuclear programme should be assessed in the context of other nuclear programmes in the Middle East, especially the Israeli programme, and that if Iran developed a nuclear capability, this would contribute to Middle East stability as it would establish a balance of terror with Israel.

These perceptions were corroborated by the findings of a number of major survey projects that were conducted across the Arab region in the post-9/11 era. The 2011 Arab Opinion Index, carried out by the Doha-based *Arab Center for Research and Policy Studies* (ACRPS) through face-to-face interviews with a total sample of 16,173 respondents in twelve Arab countries, highlighted aggregate trends in Arab public opinion on a number of regional issues, including the sources of threats to Arab security and nuclear proliferation. Taken together, these countries (Egypt, Algeria, Iraq, Jordan, Lebanon, Mauritania, Morocco, Palestine, Saudi Arabia, Sudan, Tunisia, and Yemen) represented 85 % of the total population of the Arab world. The survey found that the majority of Arabs (51 %) believe that Israel is the most significant threat to Arab national security. The United States came second on the list, as 22 % of Arabs saw it as the biggest threat to the security of Arab countries, while Iran came in at a very distant third place with 5 % of Arabs seeing it as the biggest threat to Arab security. The survey also showed that although the majority of Arabs (55 %) are generally in favour of establishing a nuclear-free zone in the Middle East, a bigger majority (69 %) agree that Israel's possession of nuclear weapons and its reluctance to disarm justifies nuclear proliferation by other countries in the region, including Iran.[42]

Arab perceptions were also confirmed by another survey project conducted by the Anwar Sadat Chair for Peace and Development at the University of Maryland under the leadership of Shibly Telhami, in collaboration with Zogby International. The Sadat Chair/Zogby International survey served as an umbrella project under which six surveys were conducted at annual intervals from 2003 to 2010. On the one hand, the survey, which covered respondents in Egypt, Jordan, Saudi Arabia, UAE, Morocco, and Lebanon, found that the majority of Arabs (78 % in 2006 and 83 % in 2008) had unfavourable views towards the United States. It also showed that the issues pertaining to the control of oil, the protection of Israel, and the domination of the Arab world ranked highest among the list of factors Arabs considered the most important in driving American policy in the Middle East. In general, the survey indicated that the majority of Arabs (69 % in 2006 and 70 % in 2008) had no confidence in the United States. On the other hand, the survey found that the majority of Arabs believed that Iran has the right to develop a nuclear programme and that international pressure on Iran to curtail its programme should stop. Indeed, the percentage of Arab approval for Iran's nuclear programme jumped from 60 % in 2004 to 67 % in 2008, although approximately half of all

42 "The Arab Opinion Project: The Arab Opinion Index". *Arab Center for Research and Policy Studies*, March 2012.

respondents in the surveys suspected that Iran's nuclear programme was intended for weapons manufacture. When asked in 2008 about the potential outcome of Iran's acquisition of nuclear weapons for the Middle East region, the majority of correspondents said that the outcome would be positive, and viewed a nuclear-armed Iran as a desirable counterweight to US and Israeli military dominance in the region. In fact, the highest approval rate was recorded in Saudi Arabia—the hotbed of Sunni activism in the Arab world—where 73 % of the population expressed positive views towards the Iranian nuclear programme.[43]

The Pew global attitudes project, measuring international opinion about the United States, reinforced such findings. In May 2007, Princeton professor Steven Kull presented the results of his survey of attitudes towards the United States in the Muslim world, including a number of Arab countries, to the House of Representatives committee on foreign affairs. The survey showed that 93 % of Egyptians and 76 % of Moroccans had an 'unfavourable' view of the Bush administration. This negative perception of the United States was also reflected in Arab views about actual US goals towards the Muslim and Arab world. The survey found that 92 % of Egyptians and 78 % of Moroccans believed that the real goal of the United States was to divide and weaken the Islamic world (Kull 2007).

In the light of this review, one could articulate two main observations about Arab perceptions of regional arms control. The first is that Arab perceptions of arms control and other related hard security issues went through major changes in the aftermath of 9/11 and the 2003 invasion of Iraq, as a result of American pressure. These changes were in the direction of placing less emphasis on the question of arms control and the Arab–Israeli conflict, which was replaced by the Arab–Iranian conflict over the questions of the dissemination of Shi'ism and the Iranian nuclear programme. In this, Arab countries had to conform to American policies under the threat of regime change. The second observation is that, contrary to conventional theory, Arab civil society is more hawkish than the ruling elites as far as its perceptions of arms control and non-proliferation issues are concerned. This should lead us to question the accuracy of the contention that interaction among civil society groups across the region would facilitate peace under the conditions of Israeli occupation of Arab territories, coupled with the existing power disequilibrium in favour of Israel. In addition, the dovishness of Arab regimes, compared with civil society groups, did not lead to any significant breakthroughs in dealing with the major hard security issues in the region, namely the Arab–Israeli conflict and Israel's monopoly of nuclear armament. This failure discredited the Arab regimes and reinforced the perceptions of civil society groups. It also led, among other factors, to the discontent of the Arab masses against their ruling elites, which eventually culminated in the outbreak of mass uprisings against a number of Arab regimes in what became widely known as the Arab Spring.

43 See the results of the 2003–2008 Anwar Sadat Chair/Zogby International surveys in the Arab world; available at: <http://sadat.umd.edu/surveys/index.htm>.

References

Abdel-Salam, M. (1992). The global control of middle eastern armament. *Journal of International Politics* (Cairo),*103* October (in Arabic).

Al-Dessouki, M. (1996). The control of armament in the Middle East. *Strategic Paper Series* (Cairo), *42* (in Arabic).

Boudreaux, R., & Daragahi, B. (2007). Israel lifts veil on air strike against Syria. *Los Angeles Times*. Retrieved October 3, 2007, from http://articles.latimes.com/2007/oct/03/world/fg-airstrike3.

Butcher, T. (2007). Syria accuses Israeli warplanes of entering territory. *Daily Telegraph*. Retrieved September 9, 2007, from http://www.telegraph.co.uk/news/main.jhtml?xml=/news/2007/09/07/wsyria107.xml.

Chirac, J. (1992). Proliferation, non-proliferation, deterrence. *Politique Internationale, 56* (Summer).

Cooper, H., & Mazzetti, M. (2007). An Israeli strike on syria kindles debate in the U.S. *New York Times*. Retrieved October 10, 2007, from http://www.nytimes.com/2007/10/10/washington/10diplo.html?ei=5088&en=d33ecba9c972245a&ex=1349668800&adxnnl=1&partner=rssnyt&emc=rss&adxnnlx=1192017815-HYCTUmQrKnsb00jUp5Wx9A.

Dean, J. (1988). Berlin in a divided Germany: An evolving international regime. In A. L. George, P. J. Farley, & A. Dallin (Eds.), *U.S.-Soviet Security Cooperation: Achievements, failures, lessons*. Oxford: Oxford University Press.

Diyab, M. (1995). Arms control and Arab–Israeli settlement: A Syrian perspective. In M. Diyab (Ed.), *Arms control and security in the Middle East: In search for common ground.* Cairo: Al-Ahram Center for Political and Strategic Studies (in Arabic).

El-Shazly, F. (2000). The development of the Euro-Mediterranean charter for peace and stability. In M. Ortega (Ed.), *The future of the Euro-Mediterranean security dialogue*. Paris: Institute for Security Studies-Western European Union.

Farley, P. J. (1988). Strategic arms control. In A. L. George, P. J. Farley, & A. Dallin (Eds.), *U.S.-Soviet Security Cooperation: Achievements, failures, lessons*. Oxford: Oxford University Press.

Hanrieder, W. F., & Auton, G. P. (1980). *The foreign policies of West Germany, France, and Britain*. New Jersey: Prentice-Hall, Inc.

Jervis, Robert (1976). *Perception and Misperception in International Politics*. Princeton: Princeton University Press.

Jones, P. (2003, November). Negotiating regional security and arms control in the Middle East: The ACRS experience and beyond. *Survival*, *14*, 31.

Karem, M. (1995). The Middle East: Existing status of regional efforts and arrangements, In T. A. Couloumbis, & T. P. Dokos (Eds.), *Arms control & security in the Middle East & the CIS republics*. Athens: Hellenic Foundation for European and Foreign Policy.

Klein, Y. (1995). Russia and a conventional arms: Non-proliferation regime in the middle east. In E. Inbar & S. Sandler (Eds.), *Middle Eastern security: Prospects for an arms control regime*. London: Frank Cass.

Kristensen, Hans M. & Handler, Joshua (1995). Nuclear Counterproliferation in the Middle East. *Middle East Report*, *25* (November/December); available at: http://www.merip.org/mer/mer197/nuclear-counterproliferation-middle-east.

Kull, Steven (2007). Negative Attitudes toward the United States in the Muslim World: Do They Matter?. *World Public Opinion*, *17* (May); available at: http://www.worldpublicopinion.org/pipa/articles/brmiddleeastnafricara/361.php?nid=&id=&pnt=361&lb=brme

Larrabee, S. F., Green, J., Lesser, I. O., & Zanini, M. (1998). *NATO's Mediterranean initiative: Policy issues and dynamics*. Washington, DC: Rand.

Maresca, J. J. (1988). Helsinki accords 1975. In A. L. George, P. J. Farley, & A. Dallin (Eds.), *U.S.–Soviet Security Cooperation: Achievements, failures, lessons*. Oxford: Oxford University Press.

Mekheimar, O. F. (2002). *Common challenges to the levant: Cooperative security in the Middle East after peace*. Rome: NATO Defense College.
Mintz, Alex and DeRouen Jr., Karl (2010). *Understanding Foreign Policy Decision Making*. Cambridge: Cambridge University Press.
Moubayed, S. (2007, September 20–26). With friends like these: America and Israel spin propaganda circles around the Arabs. *Al-Ahram Weekly* (Cairo).
Nafaa, H. (2004, January 1–7). Dubious Courage and Doddering Wisdom. *Al-Ahram Weekly* (Cairo).
Navias, M. S. (1994). Arms trade and arms control in the Middle East and North Africa since operation desert storm. In *The Middle East and North Africa*. London: Europa Publication.
Ortmayer, L. L. (1975). *Conflict, compromise, and conciliation: West German-Polish normalization 1966–1976*. Warsaw: University of Denver.
Pande, S. (1998, December). Nuclear weapon-free zone in the Middle East. *Strategic Analysis, XXII*, 9.
Said, A. M. (1995). Conventional arms control in the Middle East. *Strategic Papers Series* (Cairo), *29* (in Arabic).
Selim, Gamal M. (2011). Perceptions of Hard Security Issues in the Arab World. In Hans Günter Brauch et al., (Eds.), *Coping with Global Environmental Change, Disasters and Security – Threats, Challenges, Vulnerabilities and Risks*. Berlin – Heidelberg – New York: Springer.
Selim, M. E.-S. (2000). Towards a new WMD agenda in the Euro-Mediterranean partnership: An Arab perspective. In A. Vasconcelos, & G. Joffe (Eds.), *The Barcelona process: Building a Euro-Mediterranean regional community*. London: Frank Cass.
Selim, M. (2008, April 20). Revealing the veiled reality of arab nuclear programs. *Al-Arabi* (Cairo).
Shehata, D., & Wahid, M. (2011). The engines of change in the Arab world. *Journal of International Politics* (Cairo), *184* (April), 10–17 (in Arabic).
Spear, J. (1994, December). On the desirability and feasibility of arms transfer regime formation. *Contemporary Security Policy*, *15*, 3.
Steinberg, G. M. (1994). Middle east arms control and regional security. *Survival, 36*,1 (Spring).
Urwin, D. W. (1997). *A political history of Western Europe since 1945*. London: Longman.

Chapter 6
The Arab Spring and Arms Control: Any Change in Arab Perceptions?

The Arab world is at present undergoing one of its most profound political upheavals in decades. This is represented in the wave of the Arab Spring mass uprisings which has dominated the Arab world since 2011. The Arab Spring has sanctioned the beginning of a new chapter in Arab politics. It has broken the long-standing taboo that saw Arab leaders as immune from mass revolutions and viewed political change as a top-downwards process initiated by either military coups or foreign intervention. Indeed, the Arab Spring has signalled the first manifestation of mass popular demand for more democratic governance in the Arab world. In addition, it has already brought back participative politics to these countries. One of the main features of modern Arab politics was the death of participative politics, as the political process was monopolized by corrupt elites. After the democracy wave, the people became a central actor in the political process. Younger generations are now engaged in politics. Even Salafi fundamentalists and Sufi spiritualists, who used to stay aloof from politics, are now well-entrenched in the political process through their own political parties.

The Arab Spring created hopes that the Arab regional system would restore its strength as the impact of the Tunisian revolution was quickly echoed in Egypt, and what happened in Egypt was quickly transferred to other Arab countries. After the 1967 Arab military defeat by Israel, it was widely claimed that the Arab regional system and the notion of pan-Arabism had come to an end. As the Arabs became engulfed with the task of regaining the occupied territories, with the death of Nasser in 1970, and with newly-gained oil wealth and petrodollars with the resulting disorders in the Arab regional system, which appeared as a system with no leadership, it became conventional to argue that pan-Arabism was a matter of history. This was the main contention of Fouad Ajami in his famous article "The end of pan-Arabism", published in *Foreign Affairs* in 1978. Such contention was corroborated in the light of Egypt's unilateral peace with Israel in 1979, the Israeli invasion of Lebanon in 1982, and the Iraqi invasion of Kuwait in 1990, followed by the invasion and destruction of Iraq in 2003. Meanwhile, the non-Arab actors in the Middle East gained new influence, to the detriment of the dwindling Arab system.

G. M. Selim, *Global and Regional Approaches to Arms Control in the Middle East*,
SpringerBriefs in Environment, Security, Development and Peace 4,
DOI: 10.1007/978-3-642-29314-6_6, © The Author(s) 2013

With the outbreak of the Arab Spring uprisings, several analysts pointed to the revival of the Arab regional system as a direct possible consequence. They contended that the Arab Spring would bring major transformations in Arab foreign policies, as rulers would now have to pay more attention to the demands of public opinion and the competing political and social groups that sought to shape and harness its power. This would result in the pursuit of more independent foreign policies in the future, especially with respect to hot regional issues such as the Arab–Israeli conflict and relationships with the United States. Although the United States has not been a key focus for protesters in the Arab uprisings, the majority in the Arab world share a belief that their ruling elites have subordinated their national interests to the United States, which in turn has been detrimental to the process of political and economic development in the Arab world. As explained above, the apparent subordination of Arab regimes to the dictates of Western powers, particularly the United States, and their failure to provide a deterrent against Israel's regional hegemony helped to reinforce the discontent of Arab masses with their ruling elites. This was obvious in the case of Egypt, where the Mubarak regime worked in close coordination with Israel against the Palestinians. This took the form of taking part in the Israeli siege of the Gaza Strip, endorsing the Israeli attack on Gaza in December 2008, and selling Egyptian natural gas to Israel at prices much lower than market prices.

In this context, foreign policy choices, particularly with regard to the Arab–Israeli conflict and the United States, may be subject to extra scrutiny by emerging ruling elites. Advocates of this view already point to the changes in the Arab–Israeli balance. The Egyptian revolution, according to this view, has ended the strategic understanding between Mubarak and Israel against Iran, Syria, and Hamas. Mubarak worked closely with Israel, under American supervision, to contain Iran and to destroy Palestinian resistance movements, particularly Hamas. He also sold Egyptian natural gas to Israel at prices much lower than global prices and certainly much lower than the price of natural gas sold to Egyptians. This explains the dismay of Israel and the American Zionist lobby at the removal of Mubarak, who was described by some of Israel's leaders as "a strategic asset" (Heilbrunn 2011). The end of Mubarak's rule has already ended Egypt's pro-Israeli stand on Middle Eastern issues.[1] Egypt will not discard the peace treaty with Israel, but it will put more emphasis on reciprocity of commitments. According to Murphy (2011), former Assistant US Secretary of State, Israel "will be facing a more engaged and aggressive political response from Egypt for the lack of progress in negotiations with the Palestinians and for the deep freeze that the so called 'peace process' has gotten into these last several years". The Arab–Israeli equation will be affected from another perspective. With the removal of Mubarak, it became possible to reconcile the differences between Fatah and Hamas and to begin talks to form a new unity government, a development rejected by Israel and discouraged by the United States.

Baroud (2011), a Palestinian analyst, contended that "The Hamas–Fatah reconciliation deal is a predictable consequence in a chain of events that has signaled the

[1] "The strategic alliance between Egypt and Israel has collapsed". *Ha'aretz*. 14 February 2011; available at: http://www.alhadath-yemen.com/news11631.html.

remaking of a region". In his view, the Middle East now is "spearheaded by a powerful Arab country that is secure enough to reach out to multiple partners, among them other Arab countries, as well as Iran, Turkey and others". Kirkpatrick (2011) also wrote in the *New York Times* that "Egypt is charting a new course in its foreign policy that has already begun shaking up the established order in the Middle East, planning to open the blockaded border with Gaza and normalizing relations with two of Israel and the West's Islamist foes, Hamas and Iran". Egypt's shifts, he added, "are likely to alter the balance of power in the region, allowing Iran new access to a previously implacable foe and creating distance between itself and Israel, which has been watching the changes with some alarm." Walt (2012) shared a similar view. In his analysis of Egypt's foreign policy orientation after the revolution, he wrote:

> The Egyptian leader [Morsy] is not [...] the same sort of tame client that Hosni Mubarak was. Recall that one of the key themes of the Egyptian revolution was the desire to restore a sense of "dignity", both with respect to how individual Egyptians were treated but also with respect to Egypt's posture vis-à-vis the United States and other states. As I read it, Morsy is working to rebuild Egypt's ties in several directions, in order to maximize Egypt's freedom of movement and diplomatic options. Not only will this enhance Egypt's regional clout, it will encourage others to do more to keep Cairo happy. This approach is also likely to be popular with a lot of Egyptians, who weren't wild about their country being a supine patsy of the United States.

At a more specific level, proponents of this view also contended that the change in orientation of Arab foreign policy would have significant implications for efforts towards arms control and non-proliferation in the Middle East. This is particularly the case with Arab stances on both the Israeli and Iranian nuclear capabilities. Fahmy (2012), Egypt's former ambassador to the United States, subscribed to this view, arguing:

> The Middle East is changing, and regional and international security ramifications undoubtedly will flow from that change... We should expect that foreign policy decisions will increasingly be affected by public opinion and open debate rather than the wisdom or dictate of leadership alone. Consequently, the best regional and international approach to deal with these new and transformative circumstances should be the embrace of an unwavering commitment to the application of the rule of law, as well as the application of regional and international norms and standards by all states, without prejudice or exception. This is of particular importance in the areas of conflict resolution, arms control, and disarmament. Open, vibrant societies striving for domestic equality will also ultimately take serious issue with political double standards and regional asymmetries, even if their preference will be to redress them through peaceful, diplomatic means.

In the same vein, Aboul-Enein (2012), an Egyptian specialist in security and disarmament studies, argued that the Arab Spring would have significant implications for the questions of arms control and non-proliferation in the Middle East. In his speech to the EU Non-Proliferation and Disarmament Conference held by the International Institute for Strategic Studies in Brussels on 3–4 February 2012, Aboul-Enein asserted that Arab foreign policy would change to become more in line with domestic aspirations and to reflect popular sentiments. Accordingly, as Arab governments became more accountable to public opinion at home, security-related issues such as "Israel's non-adherence to the NPT, the establishment of a zone free of nuclear weapons in the Middle East, the unfolding of the Iran file and its complexities, and the encouragement of the peaceful use of nuclear energy" are expected to be considered seriously during the years ahead. This does not necessarily indicate an Arab tilt toward Iran, but rather a rebalancing of Arab interests and priorities. Arabs may view Iran as a geopolitical rival, but will not see themselves directly threatened by Iranian ambitions or military capabilities. In the meantime, Israel's nuclear capabilities are most likely to re-emerge as their greatest concern.

However, as anti-revolutionary forces rose to power in the Arab Spring countries, and Western powers began to intervene in these countries, the expectations about the re-emergence of the Arab regional system were drastically reduced. Indeed, the speculation that the Arab Spring countries would pursue an independent, proactive foreign policy in the wake of the mass uprisings has appeared to be more a product of wishful thinking than of objective analysis. The Arab Spring, as has recently become clear, is unlikely to bring any major transformation in the foreign policy orientation of these countries, which will remain politically introverted over the short and medium term, detracting from their capacity to influence regional politics. This suggests that Arab perceptions of the international and regional environments, including perceptions of arms control and non-proliferation in the region, will follow a pattern of continuity over the years ahead. One could outline a number of indicators to support this argument.

On the one hand, the Arab countries have been plagued since independence by the failure to achieve real indigenous development, by domination of corrupt and dictatorial military elites, and by dependency on the West. These conditions have complicated the task of radical democratic transformation, as old elites are well-entrenched in the structures of economic and political power. In fact, the revolutionaries who initiated the Arab Spring uprisings did not assume the leading positions in dismantling the old authoritarian regimes and building new democratic ones. Ironically, elements of the old regimes assumed these tasks. For the first time, we have democratic projects designed and implemented by elements who were the main actors in the old dictatorships. This strongly suggests that the pre-Arab Spring perceptions of arms control will continue to dominate the foreign policy doctrines of most Arab regimes.

On the other hand, the United States was able to contain the revolutionary wave of the Arab Spring and direct it to serve American interests in the region. The United States viewed the Arab Spring, which saw in its first wave the fall of two pro-American Arab regimes, as a direct threat to American strategic interests in the region. It was worried about losing its Arab client regimes in the Middle East and the damage that revolutionary regimes could do to American strategic interests in the region, including possible setbacks for the US-led war on terror, potential volatility in energy markets, reduced access to energy supplies, the nationalization of corporate holdings, and the endangering of the security of Israel. This was particularly true with respect to the war on terror, where the United States relied heavily on its Arab allies in the domain of intelligence gathering and the interrogation of hundreds of terrorist suspects. Indeed, Scheuer, a former officer in the US *Central Intelligence Agency* (CIA) and a special advisor to its chief from 2001 to 2004, warned that the Arab Spring represented "an intelligence disaster" for the United States and other Western powers. "The help we were getting from the Egyptian intelligence service, less so from the Tunisians but certainly from the Libyans and Lebanese, has dried up…The amount of work that has devolved on US and British services is enormous, and the result is blindness in our ability to watch what's going on among militants," he said.[2]

In response to these challenges, the United States moved quickly in the direction of containing the Arab uprisings through different strategies. In Tunisia, Egypt, and Yemen, the United States openly supported the Ben Ali, Mubarak, and Saleh regimes before and during the mass protests. However, when it became clear the dictators were collapsing, the United States changed tactics by siding with the revolutionary forces, while working, especially in the case of Egypt and Yemen, to maintain the main power structures that would serve American interests. In Libya, the United States managed to reorient the revolutionary process from non-violent resistance to an all-out war launched by the local opposition and Western powers, resulting in the destruction of the main structures of the state and its power. In other Arab countries (mainly Bahrain), the United States assisted the regimes in aborting the uprisings and crushing the nascent democratic movements before they could reach critical mass, thus reinforcing existing political orders.

In Egypt, for example, the United States sought to secure the Mubarak regime during the mass uprising of January 25. However, after it became clear that the revolution was gaining momentum, the United States changed sides and openly endorsed regime change in Egypt. It also began implementing a new strategy to influence the course of events in post-Mubarak Egypt, centred on creating an alternative client regime. In this context, the United States threw its support behind the *Supreme Council of the Armed Forces* (SCAF) which assumed power following the resignation of Mubarak. It worked through the SCAF to ensure that Egypt remained compliant with US interests in the region. The US support for the ruling Egyptian military junta was understandable as that the majority of aid sent to the country has gone directly to the armed forces—including US$1.3 billion in 2011 alone. Indeed, the Obama administration was keen to maintain this level of funding despite America's own financial troubles. This gave the United States substantial leverage over the SCAF and the power to influence its policies.

This influence was most evident in the area of foreign policy, with the ruling military junta conforming to the legacy of Mubarak, which was excessively conciliatory to the United States and lenient towards Israel. This continuity with 'Mubarakism', as one Egyptian activist described it, was manifested on several fronts (Shama 2012). The first followed immediately on the resignation of Mubarak when the SCAF issued a statement assuring the United States and Israel that the 1979 Egyptian–Israeli peace treaty would remain intact despite the downfall of Mubarak. In addition, the SCAF backtracked on what appeared to be an orientation towards the re-establishment of diplomatic relations with Iran. It also backtracked on its initial announcement that Egypt's Rafah border crossing into the Gaza Strip would remain open on a permanent basis. The SCAF even retained Mubarak's widely-criticized policy of exporting natural gas to Israel at prices much lower than world market prices. This resulted in an increasing number of sabotage attempts on

2 "Arab spring has created 'intelligence disaster', warns former CIA boss". *The Guardian*, 28 August 2011; available at: http://www.guardian.co.uk/world/2011/aug/28/arab-spring-intelligence-disaster-scheuer.

the natural gas pipeline to Israel by underground groups. In round table discussions held on 27 July 2011 at the US Institute for Peace in Washington, Major General Said Elassar, a member of the SCAF, frankly acknowledged the US influence over SCAF. Asked whether Egypt would continue with or break away from its pro-American foreign policy orientation, Elassar responded, "We have strong strategic relations with the United States since the 1979 Camp David Accords. In military to military relations, the US is our pillar. We have been supported by the United States, and we are proud to protect United States interests" (cited in Omestad 2011).

As the SCAF became the target of huge pro-democracy demonstrations, the United States sought to establish links with the conservative Muslim Brotherhood as a potential ally in Egypt after Mubarak, and to forge an understanding between the SCAF and the Brotherhood on the main elements of the transition period and the nature of the future regime. In parallel, the Brotherhood was looking to secure US support for its rise into power in Egypt. The US–Brotherhood understanding was evident on many fronts, the most noticeable was the role the Brotherhood played in releasing leaders of the American civil society groups who had been arrested and charged with violating the law by working in Egypt and receiving foreign funds without government permission. On 1 March 2012, US Senators John McCain, Lindsey Graham, John Hoeven, and Richard Blumenthal issued a statement in which they openly thanked the Muslim Brotherhood for their role in releasing these people, although such a release represented an open violation of Egyptian law.[3] In the lead-up to the June 2012 presidential elections, the Brotherhood reassured the Obama administration that US interests would be maintained in the event they reached power in Egypt.[4] The Brotherhood's reassurances included "no violation of the peace treaty with Israel, no interruption of economic cooperation between Egyptian businessmen and their Israeli counterparts, no uncalculated support to Hamas in Gaza, and no rushed relations with Iran" (Ezzat 2012). It came therefore as no surprise that the Obama administration threw its support behind the Brotherhood's presidential candidate Mohammad Morsy, a determining factor in his victory.[5]

---

[3] The statement is available at: http://www.mccain.senate.gov/public/index.cfm?FuseAction=PressOffice.PressReleases&ContentRecord_id=CF34DE54-D3B5-651D-C3EE-505C71941FF2.

[4] "Muslim Brotherhood seeks U.S. alliance as it ascends in Egypt". *The Washington Times*, 5 April 2012; available at: http://www.washingtontimes.com/news/2012/apr/5/muslim-brotherhood-seeks-us-alliance-as-it-ascends/?page=all.

[5] This support was clearly evident in the run-off elections between the Brotherhood's candidate Mohammad Morsy and ex-General Ahmad Shafiq. Following the end of vote-casting, the Muslim Brotherhood compiled preliminary election returns from nearly all polling centres, and unilaterally declared Morsy as the winner, with almost 52 % of the vote. Although the Brotherhood's move was illegal, the Obama administration demanded that the SCAF declared the election results without investigating the hundreds of appeals presented by Shafiq to the Election Commission. On 20 June 2012, US Secretary of State Clinton stated that Washington insisted that the SCAF must turn power over to the "legitimate winner" of the country's first post-Mubarak presidential elections, and not subvert constitutional authority. In fact, if the SCAF-appointed Election Commission had investigated the appeals, they would have reversed the preliminary results due to the forging of about a million ballots. However, on 24 June and under intense US pressure, the Commission declared Morsy the winner of Egypt's presidential elections.

The American support for the Brotherhood was not without a price. That price is represented in maintaining the main parameters of Egypt's foreign policy in line with American regional interests. It is in this context that the newly-elected president has shown no intention of breaking away from the legacy of Egypt's foreign policy under Mubarak. Indeed, the signs of continuity have been clearly evident with respect to Egypt's relations with Israel and the United States. Aside from some timid and vague statements made by the Islamists about restoring Egypt's leadership position in the Arab and Muslim worlds, President Morsy and senior members of his party eschewed, on several occasions, the Brotherhood's long-standing anti-American and anti-Israel stance. Morsy announced that Egypt would remain committed to all international obligations and treaties, including the peace treaty with Israel. The Brotherhood went even further than Mubarak by invoking the Quran to stress a religious obligation to adhere to the treaty.

It was also revealing that Morsy did not consider it worth raising the question of Israel's nuclear programme and weapons on his foreign policy platform as one of the challenges facing Egypt's national security. In fact, the issue has been totally abandoned in the official rhetoric of the new Egyptian regime, which moved instead in the direction of upgrading the level of security cooperation with Israel. On 9 August 2012, following a terrorist attack on Egyptian guard troops across the Rafah borders which led to the killing of 16 Egyptian soldiers, the Israeli daily *Haaretz* described security coordination between Egypt and Israel as being "at the highest level since the peace agreement was forged by the two nations". The newspaper added that "During an age in which the Muslim Brotherhood controls both houses of Egyptian Parliament, and Mohammed Morsi, one of the moment's leading figures in Egypt serves as president, the level of security cooperation between the two nations has undergone a significant upgrade, right under the nose of the Israeli and Egyptian publics."[6] Edris (2012), an Egyptian analyst, reiterated similar remarks, arguing that Egypt under Morsy has maintained a high level of security coordination with Israel, which could eventually lead to the establishment of an 'Egyptian–Israeli security partnership' in the future. This partnership, Edris warned, would be tantamount to a 'political alliance', and would mean the subordination of post-revolution Egypt to Israeli and American dictates in the region.

Indeed, the upgrading of Egyptian–Israeli security coordination went in parallel with the continuation of Egypt's economic blockade of the Gaza Strip. Although Morsy had initially promised to ease restrictions on Gaza, that promise has never materialized. Instead, Egypt has tightened the blockade. As reported by *The Economist*, President Morsy "has done more than Mubarak did to stem the flow of goods (including arms) through the tunnels that connect Gaza and the Sinai Peninsula".[7] Using similar rhetoric

---

6 "Israel-Egypt security cooperation at one of highest levels since peace deal, say officials on both sides". *Haaretz*, 9 August 2012; available at: http://www.haaretz.com/blogs/east-side-story/israel-egypt-security-cooperation-at-one-of-highest-levels-since-peace-deal-say-officials-on-both-sides.premium-1.457085.

7 "A honeymoon that wasn't: Egypt's new Islamist leaders have so far failed to embrace their Gazan brothers". *The Economist*, 29 September 2012; available at: http://www.economist.com/node/21563776.

to Mubarak's, Morsy made it clear to Hamas that if Egypt was to improve the situation in Gaza, Hamas would have to close the smuggling tunnels across the Egyptian–Gaza border, cooperate in closing down terrorist networks in the Sinai, and conciliate its Palestinian rival, Fatah, which runs the Palestinian Authority based in Ramallah on the West Bank. Further restrictions were also imposed on Gaza following the Rafah attack. The Morsy regime pointed the finger at Gaza and accused Hamas of involvement in the terrorist attack, and moved quickly to shut down the Rafah crossing as well as the underground tunnels.

In Libya, the United States moved quickly to act and to seize the moment which it had seemed to about to miss in Egypt, as it had long-standing bitter memories with Qaddafi. It took advantage of the Libyan uprising, which erupted on 17 February 2011, and strove to position itself at the heart of the crisis in an attempt to secure its interests in the oil-rich country, and to divert and block the revolutionary mass movements that only weeks earlier had toppled the US-backed regimes in Tunisia and Egypt. The US intervention strategy was implemented at two main levels.

First, the United States sought influence over Libya's *Transitional National Council* (TNC), established on 27 February 2011. The TNC, which consisted of army generals, ex-members of Qaddafi's government, and other long-time elite opposition figures, was to lead armed opposition against the Qaddafi regime. Although not elected, the TNC declared itself to be the only legitimate body representing the people of Libya, and canvassed the West for support against the Qaddafi regime. In response, the United States and other Western powers recognized the TNC as the legitimate government of Libya. They viewed the Council as a pliant government through which they could control the country's economic and geo-strategic resources in a post-Qaddafi political order. In fact, many of the TNC leaders were reported to have close connections with the United States. An obvious example is Mahmoud Jibril, who acted as the interim prime minister in the TNC, and formerly as the head of the National Planning Council and the National Economic Development Board.[8] Another example is former Libyan Colonel Khalifa Belqasim Haftar, a long-time CIA collaborator who was appointed the commander of ground forces in the Libyan rebel army.[9] .

Second, the United States instigated a military intervention in Libya under the pretext of 'humanitarian intervention'. This began with US attempts to internationalize the Libyan crisis by referring it, with the help of Qatar and other US allies in the region, to

---

8 A leaked US diplomatic cable from November 2009 written by the US ambassador to Libya, Gene Cretz, described Mr Jibril as a man who helped pave the way for the privatization of Libya's economy and welcomed American companies. "With a PhD in strategic planning from the University of Pittsburgh, Jibril is a serious interlocutor who gets the US perspective," ambassador Cretz wrote. "Head of Libyan 'Think Tank' Outlines Human Development Strategy". *The Telegraph*, 31 January 2011; available at: http://www.telegraph.co.uk/news/wikileaks-files/libya-wikileaks/8294558/HEAD-OF-LIBYAN-THINK-TANK-OUTLINES-HUMAN-DEVELOPMENT-STRATEGY.html.

9 "Taking Charge of Libya's Rebels: An In-Depth Portrait of Colonel Khalifa Haftar". *The Jamestown Foundation*, Vol. 2, No. 3, 31 March 2011; available at: http://mlm.jamestown.org/single/?tx_ttnews[tt_news]=37724&tx_ttnews[backPid]=567&no_cache=1

the UN Security Council. On 17 March 2011, the UN Security Council issued Resolution 1973 on the Libyan crisis. The Resolution called for the imposition of no-fly zones in Libya and authorized "all necessary measures to protect civilians and civilian populated areas under threat of attack in the Libyan Arab Jamahiriya, including Benghazi, while excluding a foreign occupation force of any form on any part of Libyan territory".[10] On 19 March 2011 the United States and its Western allies, working under the umbrella of NATO, initiated a campaign of air strikes on Libya under the justification of protecting civilians from attack by Qaddafi's forces. However, NATO went beyond the terms mandated by the UN resolution. Instead of preventing civilian casualties through "the immediate establishment of a ceasefire and a complete end to violence" as stipulated by Resolution 1973, NATO intervened on the side of rebel forces and unilaterally shifted the mission in the direction of regime change by force. This was explicitly stated by Amr Moussa, the then Secretary General of the Arab League, who on 20 March 2011, accused the United States of violating the UN resolution. "What is happening in Libya differs from the aim of imposing a no-fly zone, and what we want is the protection of civilians and not the bombardment of more civilians", he said.[11]

In the course of NATO's military campaign, the United States and its allies repeatedly rejected all proposals for a ceasefire or a negotiated settlement. On 10 April 2011, the Libyan government accepted a proposal by the *African Union* (AU) for an immediate ceasefire. The AU proposal also called for the unhindered delivery of humanitarian aid, the protection of foreign nationals, a dialogue between the government and rebels on a political settlement, and the suspension of NATO air strikes.[12] However, NATO and the TNC rejected the AU proposal, and the bombing campaign continued in full swing. On 15 April, the leaders of the United States, Britain, and France issued a joint statement in which they stated that "while our duty and mandate under UN Security Council Resolution 1973 is to protect civilians, and… is not to remove Qaddafi by force", it was nevertheless "unthinkable that someone who has tried to massacre his own people can play a part in their future government". The three leaders also rejected demands for an immediate ceasefire and a negotiated exit for Qaddafi.[13] This exposes as false the claim that the United States intervened in Libya for humanitarian purposes.

The US-led military intervention in Libya was a determinant factor in re-asserting America's influence and undermining the revolutionary movements, thus steering an initially popular uprising toward the direction of a US-engineered regime change.

---

10 See the text of UN resolution 1973 at: http://www.un.org/News/Press/docs/2011/sc10200.doc.htm.

11 "Arab League condemns broad bombing campaign in Libya". *The Washington Post*, 21 March 2011; available at: http://www.washingtonpost.com/wp-dyn/content/article/2011/03/20/AR2011032001965.html.

12 "Libya: Gaddafi government accepts truce plan, says Zuma". *BBC News*, 11 April 2011; available at: http://www.bbc.co.uk/news/world-africa-13029165.

13 "Libya letter by Obama, Cameron and Sarkozy: Full text". *BBC News*, 15 April 2011; online at: http://www.bbc.co.uk/news/world-africa-13090646.

This strongly implies that post-Qaddafi Libya will emerge as a strategic local client responsible for serving American interests in the region, including its non-proliferation agenda. This scenario became more evident following the assassination of Qaddafi, with the TNC's appointment of Mahmoud Jibril as the head of Libya's interim government and Hiftar as supreme commander of the new Libyan army. On 7 October 2011, US Republican senators John McCain, Lindsey Graham, Mark Kirk, and Marco Rubio co-authored an op-ed for *The Wall Street Journal*, following their visit to Libya, in which they predicted that post-Qaddafi Libya would be an important ally of the United States in the region.[14]. This led some analysts to draw a line of similarity between post-Qaddafi Libya and post-Saddam Iraq, with arguments that the US-led military intervention in Libya under the pretext of establishing security and peace, thereby justifying US long-term presence in the oil-rich country, was tantamount to the emergence of another Iraq in Libya and the expropriation of its natural resources (Salami 2011; Howaidy 2011).

In Yemen, the United States worked closely with Saudi Arabia to abort the mass uprising which broke out in February 2011 demanding the ousting of President Ali Abdullah Saleh, a close ally of the United States in the war on terror. In April 2011 and following a series of government massacres of the revolutionary forces, the Saudi-influenced GCC negotiated a power-sharing deal between the Saleh regime and the revolutionary forces, represented by the opposition *Joint Meeting Parties* (JMP). The American-backed GCC initiative aimed to preserve Saleh's military and security apparatuses, which had developed close ties with the United States over the past decade, and to end the mass anti-government protests. The initiative stipulated that Saleh hand over his powers to his deputy Abdu Rabu Mansour Hadi and form a unity government chaired by the opposition to formulate a new constitution and conducting elections. It also granted Saleh and his family immunity from charges of corruption and human rights abuses.[15]

The GCC initiative was negatively received by the revolutionary youth movements, which viewed the initiative as falling short of the comprehensive change protesters had been demanding for almost ten months.[16] However, after eight months of stalling and under American and Saudi pressure, the GCC initiative was signed on 23 November 2011. In line with the plan, Yemen held a presidential referendum in February 2012 which confirmed Acting President Hadi as Yemen's new president with a considerable majority. Meanwhile, former President Saleh, who received medical treatment in the United States following an assassination attempt, returned to Yemen for part of a ninety-day transition period between November 2011 and February 2012, as stipulated by the GCC initiative. He remained ensconced as president of the General People's Congress party, the

[14] John McCain, Lindsey Graham, Mark Kirk and Marco Rubio, "The Promise of a Pro-American Libya". *The Wall Street Journal*, 7 October 2011; available at: http://online.wsj.com/article/SB10001424052970203388804576613293623346516.html

[15] "From revolution to unsolved crisis". *Al-Ahram Weekly* (Cairo) 14–20 April, 2011.

[16] "Yemen's Imperialist Transition". *Yemen Times*, 6 February 2012; available at: http://www.yementimes.com/en/1529/opinion/322/Yemen%E2%80%99s-imperialist-transition.htm.

former ruling party now sharing power with the former opposition JMP. As a result, the youth-led movements that initiated the pro-democracy demonstrations were marginalized and displaced by forces from the pro-American old regime.

Accordingly, it is highly unlikely that Yemen's foreign policy will undergo any major change under the Hadi regime. In fact, the United States has closely supported the Hadi regime. In 2012, the Obama administration provided Yemen with a total of US$346 million in military and economic assistance, the largest sum in the history of US foreign aid to Yemen.[17] Another sign of close US–Yemeni cooperation was evident in the Obama administration's decision to expand US military operations in Yemen through stepped-up drone attacks and special operations forces on the ground. As part of this cooperation, an unknown number of US special operations forces have been on the ground, training the Yemeni military and providing direct logistical support, as well as directing drone attacks. On 16 May 2012, the *Los Angeles Times* reported that "the Obama administration's direct military role in Yemen is more extensive than previously reported and represents a deepening involvement in the nation's growing conflict." It added that "the U.S. counter-terrorism effort in Yemen has become broader than the decade-old pursuit of Al Qaeda in Pakistan".[18] A month later, President Obama acknowledged that the US military was actively engaged inside Yemen.[19]

In an interview on 7 March 2012 with the Riyadh-based *Al-Sharq Al-Awsat*, President Hadi affirmed that "Yemen would proceed with its war on the Al-Qaeda organization in the country", a clear sign of continuity in Yemen's role as an important partner in the US-led war on terror. Hadi also described his country's relations with Saudi Arabia and other pro-American Arab Gulf monarchies as 'strategic and exceptional', adding that "Yemen represents the strategic depth of these countries that are tied to Yemen with bonds of common religion, culture, good neighbourliness, and mutual interests".[20] In September 2012 and in another sign of the noticeably increasing level of US influence in post-Saleh Yemen, the United States deployed a Marine anti-terrorism unit to the capital Sanaa to protect the US embassy amid violent protests at the embassy over the depiction of the Prophet Muhammad in an American-made film.[21].

---

17 "U.S. Government Assistance to Yemen". Fact Sheet, Office of the Spokesperson, U.S. State Department, 27 September 2012: available at: http://www.state.gov/r/pa/prs/ps/2012/09/198335.htm.

18 "U.S. escalates clandestine war in Yemen". *Los Angeles Times*, 16 May 2010; available at: http://latimesblogs.latimes.com/world_now/2012/05/washington-escalation-american-clandestine-war-yemen-us-troops-.html.

19 "Presidential Letter—2012 War Powers Resolution 6-Month Report". Office of the Press Secretary, The White House, 15 June 2012.

20 "President Hadi, Yemen's New Leader, Lays out his Vision". *Al-Sharq al-Awsat*, 7 March 2010.

21 "US sends marines to Yemen embassy as turmoil spreads across Muslim world". *The Telegraph*, 14 September 2012; available at: http://www.telegraph.co.uk/news/worldnews/middleeast/yemen/9542014/US-sends-marines-to-Yemen-embassy-as-turmoil-spreads-across-Muslim-world.html

Jubran (2012), a Yemeni analyst, wrote an article in the Yemeni daily *Al-Akhbar* in which he described the US Ambassador to Yemen, Gerald Feierstein, as 'The New Dictator' of Yemen. He contended that his country has fallen under effective US trusteeship, arguing that the US ambassador "has assumed a de facto governing role in Yemen, pushing for progress but only in the manner that he deems appropriate, and which does not, of course, conflict with broader US policy in Yemen". Ayesh, editor of the Yemeni Daily *Al-Awwali*, shared the same view, arguing that the US ambassador to Yemen is acting "not just as Yemen's governor, but as a leader propelled by a transformative revolution into the country's top position". He added that "the political and military classes surrendered the country's affairs completely to the world powers, and then preoccupied themselves with their internecine struggles" (cited in Jubran 2012).

In Bahrain, home of the US Navy's Fifth Fleet, the United States launched a covert counter-revolution to crush the pro-democracy uprisings which erupted in February 2011. Despite the apparently democratic, non-sectarian intentions of the protesters, the United States and Saudi Arabia were quick to play the sectarian card and frame the conflict as one between Sunni and Shi'a groups, rather than between an oppressive regime and the disillusioned masses. They used the claims of Iranian meddling and sectarian politics to justify a military intervention in Bahrain.[22] In this context, Saudi Arabia, in coordination with the United States, used a shared defence clause in the GCC charter as a pretext for intervention. On 14 March 2011, the Saudis moved hundreds of troops into Bahrain under the umbrella of the Peninsula Shield Force, a 10,000-man military unit founded in 1984 and made up of troops from the GCC states, to crush a rapidly escalating democratic uprising. This was the first time the Force was deployed against a domestic population. Over the next few days, the Bahraini government declared a three-month state of emergency and authorized the military to take any steps necessary to restore national security. The government also staged a violent crackdown against protesters in the capital, Manama, with tanks, armoured personnel carriers and helicopters, and destroyed the Pearl Monument, which had become the Tahrir Square of the Bahraini revolution. Many demonstrators were killed, and hundreds were injured or arrested. Within a week, the peaceful protests at the heart of Bahrain's democracy movement were shattered.

In response to these developments, the Obama administration maintained its support of the monarchy, while limiting itself to advocating some improvements to human rights conditions. Indeed, the Saudi military campaign began two days after a visit by US Secretary of Defense Robert Gates to Bahrain where he reassured King Hamad bin Isa al-Khalifa that the US government stood with the Bahraini monarchy.[23] On 13 March 2011, the White House issued a soft-language statement in which

---

[22] "Blame Iran: a dangerous response to the Bahraini uprising". *The Guardian*, 20 August 2011; available at: http://www.guardian.co.uk/commentisfree/2011/aug/20/bahraini-uprising-iran.

[23] "U.S. Defense Secretary Pays Surprise Visit to Bahrain". *The Wall Street Journal*, 11 March 2011; available at: http://online.wsj.com/article/SB10001424052748703597804576194620977454958.html.

it urged the government of Bahrain "to pursue a peaceful and meaningful dialogue with the opposition rather than resorting to the use of force". It also urged "GCC partners to show restraint and respect the rights of the people of Bahrain, and to act in a way that supports dialogue instead of undermining it".[24] Using the same rhetoric, US Secretary of State Clinton urged both the Bahraini and Saudi regimes to begin negotiating a resolution of the conflict with the protesters, but other than dispatching an Assistant Secretary of State to facilitate talks, nothing further was done. In a direct acknowledgment of the Saudi intervention, US Senator John Kerry claimed that the Saudi force "was not looking for violence in the streets". Rather, "They would like to encourage the king and others to engage in reforms and a dialogue", he said.[25]

The US counter-revolution in Bahrain was further evident with the Obama administration's decision in May 2012 to resume a $53 million arms sale to Bahrain. The arms deal, which included weapons of repression such as armoured cars and tear gas, was resumed despite the Bahraini regime's systematic suppression of activist groups.[26] The Bahraini arms deal followed the conclusion in December 2011 of another, even bigger, arms deal with Saudi Arabia. Under the terms of the $30 billion agreement, Saudi Arabia would get eighty-four new F-15 jets and upgrades to another seventy F-15s in the Saudi fleet with new munitions and spare parts. Shapiro, US Assistant Secretary of State for Political-military Affairs, described the agreement as further evidence of America's determination to project its political and military influence in the Gulf region. "This sale will send a strong message to countries in the region that the United States is committed to stability in the Gulf and the broader Middle East… It will enhance Saudi Arabia's ability to deter and defend against external threats to its sovereignty," he said.[27]

It is obvious, then, that the United States was able to carry out a process of political engineering under which it reversed a revolutionary trend which could have jeopardized American interests in the Arab world. After initially supporting its autocratic allies, the United States shifted its policy towards containing the revolutionary movements and employing them to secure its interests. This took the form of striking alliances with the Muslim Brotherhood in Egypt through which the Americans would secure the Brotherhood's control of power and the latter would make sure that American interests in the Arab world would be maintained. This resulted in a domestic political course designed to facilitate the Brotherhood's road to power, which created deep divisions in Egypt that are still

---

[24] "Statement from the Press Secretary on violence in Yemen and Bahrain". The White House, Office of the Press Secretary, 13 March 2011.

[25] "The Proxy Battle in Bahrain". *The New York Times*, 19 March 2011; available at: http://www.nytimes.com/2011/03/20/weekinreview/20proxy.html?pagewanted=all.

[26] "US resumes arms sales to Bahrain. Activists feel abandoned". *The Christian Science Monitor*, 14 May 2012; available at: http://www.csmonitor.com/World/Middle-East/2012/0514/US-resumes-arms-sales-to-Bahrain.-Activists-feel-abandoned.

[27] "With $30 Billion Arms Deal, U.S. Bolsters Saudi Ties". *The New York Times*, 29 December 2011; available at: http://www.nytimes.com/2011/12/30/world/middleeast/with-30-billion-arms-deal-united-states-bolsters-ties-to-saudi-arabia.html.

plaguing the country so far. In the case of Libya, the United States intervened militarily to destroy the infrastructure of the state and kill its leader, and this resulted in a wave of violence and the fragmentation of the state. Also, the United States intervened in the Yemeni case to secure the rule of Saleh, who was a close ally of the USA in its "war on terror." This resulted in prolonged turmoil in Yemen which has almost wrecked the country. In the meantime, as the Arab Spring countries got busy with their domestic turmoil, the GCC countries began to assume leadership of the regional system and use it to suppress revolutionary changes. This was articulated in the roles the GCC played in suppressing the Bahraini reform movement and inviting foreign intervention in Libya and Syria under the pretext of supporting the revolution. The GCC states, particularly Saudi Arabia and Qatar, took an active part in supporting the anti-regime forces, not to help the peoples to gain their freedom, but in order to achieve an American agenda.

In this context, one could talk, as explained by Cavell (2012), of the "continuity of US hegemony" in the Middle East in the aftermath of the Arab Spring. Higazy (2012), an Egyptian scholar, also wondered if we should say "goodbye to the Arab Spring", arguing that the Arab revolutions were hijacked by the United States and its close allies in the region, mainly Saudi Arabia and Israel. Hence, it remains unclear, whether a more democratic Arab world could emerge from the ongoing uprisings and political transitions. The United States managed to block and divert the revolutionary wave of Arab mass uprisings, and this in turn has complicated the task of a genuine democratic transition in the Arab world. Under these conditions, it becomes even more difficult to expect the rise of a new post-revolutionary Arab foreign policy that is not subordinate to US dictates and interests. Rather, one can expect to see a pattern of continuity in Arab foreign policies, since most Arab countries, including the Arab Spring ones, will remain tightly connected to the US–Israeli regional agenda. This is particularly the case in the area of hard security issues such as arms control and non-proliferation, where Arab regimes will most likely continue to subscribe to the US agenda in an attempt to secure American support in the midst of volatile internal conditions. This might take the form of (i) downplaying the threat of Israel's nuclear weapons (ii) supporting the US sanctions regime on Iran, and (iii) supporting, whether directly or indirectly, a possible US-led military action against Iran to destroy its nuclear facilities.

References

Aboul-Enein, S. (2012). Non-proliferation and security in the Middle East. A speech delivered to the EU Non-Proliferation and Disarmament Conference. The International Institute for Strategic Studies, Brussels (3–4 February). Retrieved from http://www.nonproliferation.eu/documents/firstconference/samehaboulenein.pdf.

Baroud, R. (2011, May 5–11). The New Middle East. *Al-Ahram Weekly* (Cairo), *1046*.

Cavell, C. S. (2012). From the Arab spring to the Islamic awakening: Continuity of US hegemony. *Global Research*. Retrieved 16 April, 2012, from http://www.globalresearch.ca/index.php?context=va&aid=30178.

Edris, M. S. (2012). The Rafah crime and strategic partnership with Israel. *Al-Ahram* (Cairo), 13 August (in Arabic).

Ezzat, D. (2012, July 17). Road towards US–Brotherhood friendship trodden long before Morsi's election. *Al-Ahram* (Cairo).

Fahmy, N. (2012). The Arab awakening and regional security. *The Nonproliferation Review*. Retrieved June 14, 2012, from http://wmdjunction.com/120614_arab_awakening.htm.

Heilbrunn, J. (2011). Israel and the Arab revolutions. *The National Interest* (14 February). Retrieved February 14, 2011, from http://nationalinterest.org/blog/jacob-heilbrunn/israel-the-arab-revolutions-4872.

Higazy, A. (2012, September 1). Shall we say goodbye to the Arab spring's uprisings? *Al-Ahram* (Cairo).

Howaidy, F. (2011, September 27). On the kidnapping of the Arab spring. *Al-Shorouk* (Cairo), (in Arabic).

Jubran, J. (2012, July 10). US ambassador in Yemen: The new dictator. *Al-Akhbar* (Sanna).

Kirkpatrick, D. D. (2011). In shift, Egypt warms to Iran and Hamas, Israel's foes. *The New York Times*. Retrieved April 28, 2011, from http://www.nytimes.com/2011/04/29/world/middleeast/29egypt.html?_r=0.

Murphy, R. (2011). New chapter in the Middle East. Council on Foreign Relations. Retrieved March 3, 2011, from http://www.cfr.org/middle-east/new-chapter-middle-east/p24276.

Omestad, T. (2011). Beyond Tahrir: The trajectory of Egypt's transition. United States Institute of Peace. Retrieved July 27, 2011, from http://www.usip.org/publications/beyond-tahrir-the-trajectory-egypts-transition.

Salami, I. (2011). Libya after gaddafi. *Foreign Policy Journal*. Retrieved August 25, 2011, from http://www.foreignpolicyjournal.com/2011/08/25/libya-after-gaddafi/.

Shama, N. (2012, February 7). Decoding Egypt: The dominance of Mubarakism. *The Daily News Egypt*.

Walt, S. M. (2012). Friedman's sermon. *Foreign Policy*. Retrieved August 29, 2012, from http://walt.foreignpolicy.com/posts/2012/08/29/friedmans_sermon.

Chapter 7
Conclusion and Policy Recommendations

The global arms control approach in the Middle East has followed a pattern of continuity ever since the end of the Cold War. During the 1990s, the global arms control proposals in the Middle East were built on the assumption that an Israeli monopoly of nuclear weapons would serve Western interests in the region. These proposals, which were mainly initiated by Western powers, advocated the removal of chemical and biological weapons from all Middle Eastern countries, but restricted the removal of nuclear weapons to Arab countries and Iran. In the field of nuclear weapons, the Western proposals emphasized a CBMs approach rather than a comprehensive nuclear ban approach, equivalent to the maintaining of the nuclear status quo in the form of an Israeli nuclear hegemony in the region. This period was characterized by the existence of significant differences between the Arabs and Israel on issues relating to the very basic conceptual underpinnings of arms control. Indeed, the positions of both parties were almost zero-sum, which, in combination with the West's double-standards approach, undermined the prospects for reaching a genuine arms control regime in the Middle East.

The Western selective approach to arms control in the Middle East became further evident in the post-9/11 era, when the question of regime-oriented arms control was abandoned in favour of coercive unilateral disarmament. This shift was basically an outcome of the Bush doctrine which downplayed the significance of deterrence and multilateral arms control regimes as incompatible with the new post-9/11 security challenges, and centred, instead, on pre-emption and muscular counter-proliferation. In fact, the Bush doctrine almost returned the Arab world to a 'Hobbesian' state of nature in which hard power became the main tool of Western powers, particularly the United States, in achieving their interests in the region, including the counter-proliferation of WMD. The United States launched the 'war on terror' and employed military instruments, including military invasions, to eliminate its opponents under the pretext of counter-terrorism. It also focused on eliminating WMD from Arab and Iranian hands, while leaving those with Israel intact. Under such existential pressures, the majority of Arab regimes brought their

G. M. Selim, *Global and Regional Approaches to Arms Control in the Middle East*, SpringerBriefs in Environment, Security, Development and Peace 4,
DOI: 10.1007/978-3-642-29314-6_7, © The Author(s) 2013

foreign policies in line with the American–Israeli WMD agenda in the Middle East, as the cost of disagreement with that agenda was extremely high.

At present, the prospects for progress on arms control in the Middle East are quite poor. With no end in sight to the Arab–Israeli conflict, with increasing regional uncertainties caused by the Arab uprisings, and with talk of possible military action by Israel and/or the United States against Iran to halt or destroy its nuclear programme, the prospect of the main regional actors cooperating with each other and placing mutual and verifiable limitations on their national sovereignty and defence armaments for the collective goal of promoting regional security seems highly unthinkable at the present time.

On the one hand, it seems that the United States is quite content with the success it has achieved in containing the mass uprisings and avoiding the rise of anti-American regimes in the Arab Spring countries. When the Arab Spring broke out in Egypt and Yemen, the United States initially tried to keep their proxy rulers in power as long as possible. However, once persistent uprisings made the continued rule of these loyal autocrats untenable, the United States changed tactics by forcing them, through political and military pressures, to leave the political scene, while preserving the two respective client states throughout carefully-designed strategies that brought new patron ruling elites into power. In the case of Libya, a complete reshuffling process took place, where the revolutionary movements were marginalized and a new pro-American regime was installed, thus replacing the highly unpredictable regime of Qaddafi. In the case of Bahrain, the United States was able to suppress revolutionary changes through covert military intervention. The self-congratulatory image of the installation of new client regimes in the Arab Spring countries, which have already become part of a comfortable strategic arrangement securing US interests in the region, will put no pressure on American policymakers to abandon the counter-proliferation agenda and revive multilateral arms control efforts in the region. In the meantime, it is highly unlikely that Arab regimes will attempt to push seriously for multilateral regional arms control because of their obsession with domestic conditions, and in order not to antagonize the United States.

On the other hand, it is unlikely that Israel and Iran will give us their nuclear ambitions, with the first being an undeclared nuclear weapon state outside the NPT, and the second being a potential nuclear weapon state in non-compliance with its NPT obligations. Whereas the former is largely a status quo issue that has long been an obstacle to establishing a zone free of nuclear weapons in the Middle East, the latter is a growing threat that further undermines non-proliferation efforts in the region. In fact, Iran's nuclear programme continues to progress despite several UN Security Council resolutions calling for its suspension and the implementation of an ever-increasing package of sanctions. Meanwhile, the United States and Israel have repeatedly warned that they might launch a pre-emptive attack against Iranian nuclear facilities. While such an attack would slow Iran's nuclear programme in the short term, it would likely push Iran to aggressively and overtly pursue nuclear weapons, which would be a major setback to non-proliferation in the region.

However, the failure of past attempts to launch a genuine arms control process in the Middle East does not mean giving up the entire concept of arms control in the region. Rather, it means that the necessary preconditions for the installation of an arms control regime were either missing or overlooked by the major global and regional actors involved. What follows is that a new strategy for addressing the question of arms control is necessary if a viable arms control process is to be launched in the region. This strategy is based on a number of conditions that must be met prior to, or in parallel with, the beginning of regional arms control talks.

The first condition is the resolution of the Arab–Israeli conflict. The Middle Eastern arms race is an outcome rather than an initial cause of conflicts. Accordingly, any attempt to control the arms race without reference to regional conflicts will mean putting the cart before the horse. It is not logical to call upon Israel to give up its nuclear deterrent while it is still in a state of war with many Arab countries. At the same time, it is not practical to ask the Arabs to destroy their chemical and biological capabilities while some of their territories are still occupied by Israel. This is not to suggest that arms control should wait until the resolution of the Arab–Israeli conflict and the establishment of peace. Rather, arms control should be approached in conjunction with the peace process. Agreements on the control of WMD must be entrenched in the peace treaties. In this case, the arms control agreements would provide security guarantees and would serve to strengthen the peace process.

The second condition is the comprehensiveness of the arms control regime. In light of the regional power asymmetries, any proposed arms control regime must deal with two main dimensions of the armaments question, namely (i) WMD and conventional weapons, and (ii) arms imports and the local arms production. On the one hand, the barrier between WMD and conventional weapons is rapidly eroding. New conventional weapons are expected to have a revolutionary impact on the future battlefield. Removing WMD without regard to the conventional balance of power is not likely to result in stability if that balance is in a state of disequilibrium. One may recall that during the 1950s Western nuclear superiority represented the main deterrent against Soviet conventional superiority in the European theatre. It was inconceivable to remove Western nuclear weapons and leave the Soviets as the only conventional power in Europe. Likewise, removing WMD from the Middle East leaving Israel's qualitatively superior conventional arsenal intact will only result in the reinforcement of disequilibrium and hence instability. Approaches to remove WMD should then proceed in parallel with approaches to control conventional weapons. On the other hand, most proposals on arms control in the Middle East have approached the issue from the angle of restricting arms 'transfers'. As the Arabs import most of their weapons while Israel and, to a lesser degree, Iran manufacture most of their arms needs, such control will only help to reinforce regional military disequilibrium. Hence, controls on the transfers of WMD ingredients should proceed in parallel with controls on the local production of those materials.

Third, an arms control regime must avoid narrowing down arms control and non-proliferation to certain actors in the region, while keeping the haves intact and privileged above the have-nots. This is achieved by including all regional actors whose capabilities directly influence Middle Eastern conflicts. As long as the global powers adopt a selective arms control and non-proliferation approach that allows a particular regional power, namely Israel, to assert the security value and military utility of its WMD, this will undermine the security incentive necessary to create a stable, sustainable non-proliferation regime.

The fourth condition is the balance of power between the concerned actors. An effective arms control process must be linked to a strategic equilibrium in the Middle East. In the absence of such equilibrium, it is extremely difficult to remove or control the proliferation of WMD. On the one hand, countries with WMD capabilities will not feel obliged to give up their weapons, and thus lose their strategic superiority in a regional environment characterized by conflict and security dilemmas. On the other hand, countries with no WMDs will always strive, under the perception of security threats, to possess WMD in order to protect their national security, and to engage positively with others who already possess such types of weapons. One may recall that in the war on terror the United States invaded and occupied two non-nuclear states, Afghanistan and Iraq, but did not invade North Korea, which already has nuclear weapons, or Iran, which could possibly acquire them in the near future. This is not to suggest that arms control should wait until the Arab countries achieve strategic parity with Israel. However, if an arms control process is to start within a situation of military imbalance (like the Middle Eastern case), it must at least seek to correct this imbalance in order to create a situation of mutual interest to the parties involved.

This could be the case, for example, with CBMs. If previous CBM proposals achieved no tangible results in the Middle East, this is because they were modelled on European CBMs whose express purpose was to stabilize the geopolitical environment that had emerged in Europe by the early 1970s. In the Middle East, the need is for measures not to stabilize the status quo, but rather to alter it, to create a dynamic for change, and to move towards conflict resolution. In this respect, a re-conceptualization of CBMs is necessary in order to adapt the concept to the geopolitical conditions of the region. This entails the invention of new CBMs whose task is not to consolidate existing matters in place, but rather to facilitate their transformation, thus replacing an undesirable situation with one that is desirable and satisfactory to all actors of the region. An example of such change-oriented CBMs would be the simultaneous adherence of Israel, Arabs, and Iran to the NPT, the CWC, and the BWC. This measure could have the potential to resolve the Middle East's security dilemma by removing threat perceptions among the key regional actors.

Asked about the requirements for arms control and non-proliferation in the Middle East, Al-Baradei, former IAEA Director General, emphasized the need for "a renewed security dialogue that covers all topics relevant to security in the region". This dialogue should "deal with the present security imbalance… proceed in parallel with the peace process and with the participation of all parties concerned". The dialogue should have the ultimate objective of "freeing the region of all weapons of mass destruction, including nuclear weapons... [and] limiting conventional weapons and putting in place effective confidence-building measures".[1]

Last but not least, it is crucial to address the supply side of arms control. The globalization of the arms industry, the flooding of the global arms market, and the loosening of supplier constraints have all represented major challenges to arms control efforts in the Middle East. Accordingly, a genuine Middle Eastern arms control regime will not be feasible without a firm commitment from great powers to arms control in the region. Despite their divergent national interests in the region, the United States, the EU, Russia, and possibly China will have to be the guardians of the arms control process. Indeed, the role of external powers will be indispensible to the success of arms control in the Middle East as they are the ones who can offer concrete security assurances to regional actors, including their closest allies, and constrain arms sales to the region. Without such commitment, the prospects for a viable arms control regime will be minimal. This is particularly the case with structural arms control, which tends to be an outsider-oriented process in a region that basically depends on the importation of weapons systems.

1 "Nuclear dynamics". *Al-Ahram Weekly* (Cairo), Issue No. 789, 6–12 April 2006.

Port Said University, Port Said, Egypt

Port Said University is a public university located in the city of Port Said at the intersection of the Mediterranean Sea and the northern mouth of the Suez Canal in Egypt. The university began operations in 1976 as the Port Said branch of Suez Canal University. In accordance with a Presidential Decree issued in February 2010, this branch became a separate university under the name of Port Said University. The university currently has ten faculties and institutes offering undergraduate, master's, doctoral and graduate diploma programmes, including the Faculties of Engineering, Commerce, Arts, Education, Physical Education, Specialized Education, Nursing, Science, Kindergarten, and the Higher Institute for Management and Computing. This is in addition to twenty-five research centres in a variety of fields, such as the Centre for Developing University Education, the Centre for Quality Control, and the Centre for Faculty and Leadership Development. Today, the university serves nearly 14,000 students in twenty undergraduate, graduate academic, and professional degree programmes, and provides the community with a variety of lifelong learning opportunities. It also hosts the Egyptian Olympic Village where various sporting activities take place.

The author is associated with the Faculty of Commerce, one of the oldest faculties of the university. The Faculty has departments of Political Science, Economics, Business Management, Statistics, and Accounting.

G. M. Selim, *Global and Regional Approaches to Arms Control in the Middle East*,
SpringerBriefs in Environment, Security, Development and Peace 4,
DOI: 10.1007/978-3-642-29314-6, © The Author(s) 2013

Biography of the Author

Gamal M. Selim is an Assistant Professor of Political Science at Port Said University (Egypt) and the British University in Egypt. He obtained a Ph.D. in Political Science from the University of Calgary (Canada). His research interests include International Relations Theory, Arms Control and Non-proliferation, Politics of Development, Democratization and Middle East Politics. He has given numerous conference papers and has published: "The United States and the Arab Spring: The Dynamics of Political Engineering", in: *Arab Studies Quarterly*, 35, 3 (Summer 2013); "Sectarianism and External Intervention: The Impact of the US Invasion of Iraq on the Rise of 'Political Sectarianism' in the Arab World"; in: *Strategic Papers Series*, (Cairo: Ahram Center for Political & Strategic Studies, forthcoming); "The Impact of Post-Saddam Iraq on the Cause of Democratization in the Arab World"; in: *International Journal of Contemporary Iraqi Studies*, 6,1 (April 2012); "Égypte: une révolution permanente, trahie ou kidnappée?", in: Khader Bichara, (dir.), *Le «printemps arabe»: un premier bilan* (Paris: Alternatives Sud, 2012); "Continuity and Change in the US Arms Control Policy in the Middle East", in: *Journal of South Asian and Middle Eastern Studies*, 35,1 (Fall 2011); "Perceptions of Hard Security Issues in the Arab World", in: Brauch, Hans Günter; Oswald Spring, Úrsula; Mesjasz, Czeslaw; Grin, John; Kameri-Mbote, Patricia; Chourou, Béchir; Dunay, Pal; Birkmann, Jörn (Eds.), 2011: *Coping with Global Environmental Change, Disasters and Security—Threats, Challenges, Vulnerabilities and Risks* (Berlin—Heidelberg—New York: Springer).

Contact: Department of Political Science, Port Said University, Port Said, Egypt 42526.

E-mail: gmselim@gmail.com.

G. M. Selim, *Global and Regional Approaches to Arms Control in the Middle East*, SpringerBriefs in Environment, Security, Development and Peace 4,
DOI: 10.1007/978-3-642-29314-6, © The Author(s) 2013

MIX
Papier aus verantwortungsvollen Quellen
Paper from responsible sources
FSC® C105338

If you have any concerns about our products,
you can contact us on
ProductSafety@springernature.com

In case Publisher is established outside the EU,
the EU authorized representative is:
Springer Nature Customer Service Center GmbH
Europaplatz 3, 69115 Heidelberg, Germany

Printed by Libri Plureos GmbH
in Hamburg, Germany